Emergency Planning and Management

Conference Organizing Committee

A Denton (Chair)
Noble Denton International Limited

R Ashford
Consultant

F Crawley
Consultant

M Evans
Royal Military College of Science

D Eves
Health and Safety Executive

N Jones
University of Liverpool

J Lakey
Consultant

J Strutt
Cranfield University

IMechE
Conference Transactions

Conference on

Emergency Planning and Management

21–22 November 1995

Organized by the
Institution of Mechanical Engineeers

In association with
the Hazards Forum

Co-sponsored by:

The Emergency Planning Society
Health and Safety Executive
Institution of Electrical Engineers
Institute of Fire Safety
Institution of Chemical Engineers
Institution of Civil Engineers
Institution of Nuclear Engineers
National Design Consultancy
Royal Aeronautical Society

IMechE Conference Transaction 1995 - 6

Published by Mechanical Engineering Publications Limited for
The Institution of Mechanical Engineers, London.

First Published 1995

ISSN 1356-1448
ISBN 0 85298 954 7

A CIP catalogue record for this book is available from the British Library.

Printed in Great Britain by Antony Rowe Ltd, Chippenham, Wiltshire

Contents

C507/001/95

Management systems, behaviour, and emergencies

A E WARING PhD, LRSC, FRSH, FIOSH (RSP) MIMgt
Alan Waring & Associates, Epping, UK

SYNOPSIS

Organisations (i.e. people) behave characteristically, not ideally. Emphases on those parts of safety management systems which address emergencies will need to shift from reactive approaches which focus on emergency procedures (the 'programmable robots' approach) to more comprehensive and pro-active approaches which focus on prevention, control, mitigation and recovery and on human factors which affect the whole system. Human factors need to be addressed not only at the level of the individual seeking to cope during an emergency but also at the overall cultural level of the organisation which affects attitudes and responses generally towards safety.

1 INTRODUCTION

Much of the legislation, standards and guidance directed at improvements in the management of health and safety, whether in the high hazard/high risk industries or elsewhere, carries an implicit assumption that organisations will respond as required.

However, experience suggests that organisations i.e. people rarely behave ideally. Rather, they behave each in their own characteristic non-ideal way which we loosely call their 'culture' (1,2,3,4,5,6). For effective emergency planning and management, therefore, it is necessary not only to have soundly based formal systems seeking to achieve best outcomes but also to address realistically the range of human factors which affect what actually happens. This kind of approach is called 'holistic' (5,6,7,8,9) as it seeks to address the totality of relevant factors and how they interact.

In seeking effective planning and management of emergencies, three requirements need to be met:

Prevention of emergencies

Control of emergency events to stop escalation and mitigate the consequences

Revovery from the emergency

The remainder of this paper will consider the first two of these requirements in outline in relation to formal management systems and human factors. Following papers examine in more detail particular aspects of these requirements.

2 SYSTEMATIC FRAMEWORK

2.1 A consensus

The study and application of systems ideas in themselves is controversial as are the notions of 'management systems' and 'safety management systems' (5,7,9). However one perceives 'a system', for the purposes of this paper, it will be assumed that in practice any organisation will require a systematic framework to ensure that (health and) safety is managed well. Such a framework, generally referred to as a 'safety management system' or SMS, should have the following main components (5, 10, 11, 12):

Policy, objectives and strategy
Organising, planning, resourcing
Implementation
Monitoring
Audits
Reviews

Of course, under these broad headings come a large array of sub-headings covering important topics such as hazard identification, risk assessment, control measures, training, and so on. Many of these will be addressed by other speakers, particularly in relation to emergency preparedness.

An SMS would need to include a specific element to address potential emergencies. Such a sub-system might be called an 'emergency management system'. Reflecting the recursive nature of systems, an emergency management system would also need to incorporate the main types of element as for the SMS overall i.e. policy, strategy, organising, planning etc. See section 3 below.

2.2 Management systems standards

A number of British and International Standards have been drawn up to cover the functionally related management areas of safety, quality and environment (SQE). Confusion has arisen about such standards and their status. There are also frequently unrealistic expectations of a 'clockworks' cause-effect relationship between applying such standards and achieving rapid success (5,9).

BS EN ISO 9000 Specification for Quality Management Systems and ISO 14001/BS 7750 Specification for Environmental Management Systems are both specification standards which carry with them third party certification schemes. However, the relationship between certification and actual quality and environmental performance in a particular organisation is very debatable. The models and approaches incorporated are over-simplified and naive as paradigms for success in this subject area.

Widespread concerns about the weaknesses of certification standards for safety management have been raised, especially if compliance certificates would encourage false inferences about full legal compliance and absence of hazards. The proposed British Standard for SMS (12) based on HS(G) 65 from the HSE (10) has therefore been drafted as a guide and not a specification.

2.3 Integrated Management Systems

Integration of safety, environmental and quality management systems is now high on the agendas of many large organisations. In principle, SQE integration sounds an attractive proposition in terms of greater efficiency and consistency. Certainly, integration of safety and environmental management systems would be an advantage for management of emergencies. For example, a toxic release is likely to have both safety and environmental impacts. However, although the goal is seductive and the rhetoric easy, in many organisations people are struggling to cope with what for them are often unfamiliar concepts in three technical specialisms - S, Q, E - all at the same time. There is also the risk of SQE managers from a quality background failing to appreciate the technical breadth, depth and legal requirements essential for adequate management of safety and environment.

Administrative integration may be straightforward but it is well known from management research and experience that 'big bang' approaches to major organisational change are more likely to fail than incremental approaches (5). Integration should therefore be a long-term goal approached gradually through better coordination between the three functions.

2.4 A holistic approach

It would be unrealistic to regard a safety management system of a particular organisation as functioning in isolation from the inner and outer contexts of the organisation (5). A holistic view of an SMS is therefore needed which considers these factors. This is no less so in focussing on emergencies as a specific aspect of the SMS. Table 1 overleaf lists key elements (5).

These contextual factors are highly relevant to changes in or affecting the organisation and which may affect safety and hence emergency considerations. They include human behaviour and 'human factors' both inside and outside the organisation. So called 'management of change' relates to changes in such contexts.

Table 1 Factors in the inner and outer contexts of an organisation relevant to SMS and emergency management

Context	Factors
Inner context	Business policy, goals and strategy Organisational structures Decision-making Nature of activities and technology used Resources Organisational history Organisational culture Power relations Risk perceptions Motivations and meanings of success
Outer context	Public policy and legislation Quality and other standards Regulatory enforcement The courts Professional and learned bodies Public opinion and the media Technology Suppliers of goods and services National and local cultures Operating terrain, physical conditions, climate Trade union policy The insurance market General trading conditions Industry sector, product markets and labour markets Shareholders

(Note: the factors are not listed in any order)

Fig 1 overleaf represents an overview model of how the author perceives the systematic elements of an SMS. This model is based on that of HS(G)65 and draft 'BS 8750' (10,12) with a few significant changes. Organising, planning and resourcing are conceived of as linked processes of preparation. Reviews emphasise the continual scanning of both the SMS and the inner and outer contexts in order to detect relevant changes.

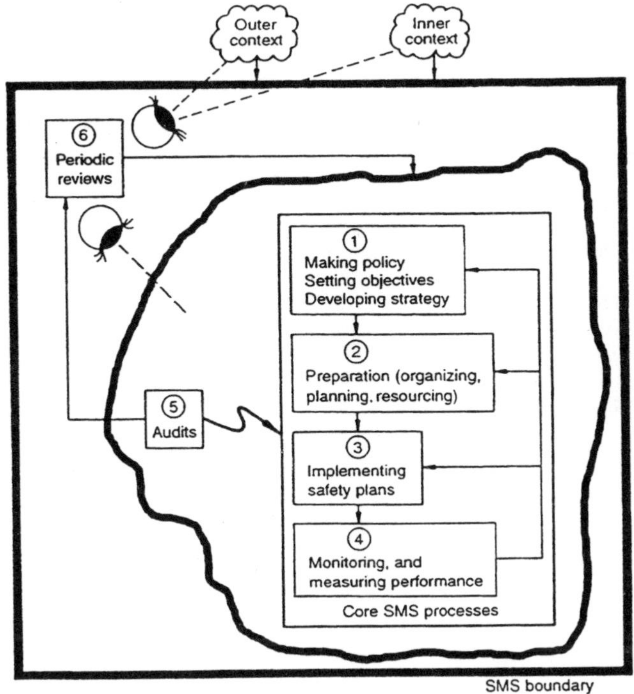

Fig 1 Overview model of systematic elements of an SMS (5)

3 KEY EMERGENCY MANAGEMENT SYSTEM REQUIREMENTS

This section of the paper merely seeks to raise some important practical considerations of an emergency management (sub)system based on the preceding description of SMS requirements. The points raised are illustrative only. Following speakers will be expanding on many of these points and also many others.

3.1 Policy, objectives and strategy

What is the policy on and objectives regarding:

command, control, coordination, authority, responsibilities?
emergency control measures?
evacuation?
training, exercises, simulations?
collaboration with emergency services?

What is the strategy and under what contingencies:

contain?
fight?
abandon?

3.2 Organising, planning, resourcing

What major hazard events and emergency scenarios have been identified?
What detailed analyses of hazards and assessments of risk have been done?
What are the organising, planning and resourcing implications of the assessments?
Who will have key roles and responsibilities?

3.3 Methods for prevention, control and mitigation

What engineering methods will be used?
What organisational controls will be instituted?
What procedures will be implemented?
What behavioural control measures will be employed?
What forms of personal protection will be required?
What combinations of methods will be used for best effect?
What assumptions are being made and are they valid?

3.4 Monitoring and testing

How will the efficacy of emergency procedures and plans be judged?
What levels and kinds of monitoring would be appropriate?
What levels and kinds of testing would be appropriate?
What balance of drills, exercises and simulations?
How will lessons learned be reported and disseminated?

4 HUMAN FACTORS ISSUES

As noted above, a formal SMS is necessary to provide a common framework for practical action. However, such a framework alone is not sufficient to ensure 'success' in relation to safety generally or emergencies specifically. Relevant human factors in the inner context including organisational culture, power relations, risk perceptions, motivations and meanings of success are pivotal in deciding what actually happens (1, 2, 3, 4, 5, 12, 13). Many of the major disasters such as Kings Cross, Piper Alpha, Clapham and Chernobyl and 'near disasters' such as Three Mile Island, bear witness to such an analysis. Such factors materially affect safety through motivations, attitudes, decision-making, allocation of resources, beliefs about cause-and-effect, and so on. The interaction between individual and group human factors should be recognised. The effects of such factors on thinking and action lie behind what Toft (6) has called 'the failure of hindsight' which continues to create 'disasters waiting to happen'.

An understanding of the prevailing set of human factors relating to the inner context is therefore important to effective management of safety, including emergencies (5, 14, 15, 16, 17). However, the traditional engineering and ergonomics view of human factors either ignores these factors or seeks to reduce them to a form which fits the assumptions of engineering or ergonomics (see for example 18). Emphasis has been on individual behaviour during a rare emergency event whereas a more balanced examination would

Relate to Ch Turn

also consider the wider, more prevalent behaviour during non-emergency conditions.

Major hazards work including emergency preparedness necessarily involves an emphasis on safety engineering. It is unsurprising that such work is dominated by engineers of various kinds. However, this can lead to a narrow view of human factors and the expectation that people's behaviour is readily predictable and controllable i.e. like 'programmable robots' or 'clockworks'.

There is now a need for more input into major hazards work from those qualified in appropriate branches of management and social sciences who can contribute a more holistic and detailed view of human factors and how undesirable characteristics may be best modified. No discipline dare claim a monopoly of wisdom in these matters and it is heartening to note the positive lead of the Hazards Forum in encouraging dialogue between engineers and other disciplines.

5 CONCLUSIONS

The emphasis of any safety management system must be on prevention, including prevention of emergencies. The SMS therefore needs to address emergency preparedness as a specific requirement for prevention, control, mitigation and recovery. In developing the SMS and emergency arrangements, there will be a need to think holistically as well as systematically and consider the inner and outer contexts of the organisation. Human behaviour, both prevalent and during emergencies, needs to be understood and engineering assumptions about complex human behaviour should be treated with caution.

REFERENCES

(1) DOUGLAS M. Risk and Blame - Essays in Cultural Theory, 1992, (Routledge, London)

(2) TOFT B. Changing a safety culture: a holistic approach, Management into the 21st Century, Bradford University, 14-16 September 1992, (British Academy of Management 6th Annual Conference)

(3) TURNER B.A. et al Safety culture: its importance in future risk management, Second Workshop on Safety Control and Risk Management, Karlstad, Sweden, 6-9 November 1989, (World Bank)

(4) WARING A.E. Organisational culture, management and safety, Management into the 21st century, Bradford University, 14-16 September 1992, (British Academy of Management 6th Annual Conference)

(5) WARING A.E. Safety Management Systems, 1995 in press,

(Chapman & Hall, London)

(6) TOFT B. and REYNOLDS S. Learning from Disasters: a Management Approach, 1994, (Butterworth Heinemann, London)

(7) CHECKLAND P. and SCHOLES J. Soft Systems Methodology in Action, 1990, (John Wiley & Sons, Chichester)

(8) WARING A.E. Systems Methods for Managers: a Practical Guide, 1989, (Blackwell Scientific, Oxford)

(9) WARING A.E. Practical Systems Thinking, 1996 in press, (Chapman & Hall/International Thomson Business Press, London)

(10) HEALTH & SAFETY EXECUTIVE Successful Health and Safety Management HS(G) 65, 1991, (HSE Books, Sudbury)

(11) HEALTH & SAFETY COMMISSION Management of Health and Safety at Work Regulations 1992, Approved Code of Practice, 1992, (HSE Books, Sudbury)

(12) BRITISH STANDARDS Guide to Occupational Health and Safety Management Systems in draft, proposed 'BS 8750', 1996 forthcoming, (BSI Standards, London)

(13) WARING A.E. Safety audits and their role in disaster prevention, International conference on emergency planning, Lancaster University, September 1991.

(14) WARING A.E. Power and culture - their implications for safety cases and EER, European seminar on human factors in emergency response offshore, Aberdeen, 29-30 September 1993, (Business Seminars International)

(15) BIGNELL V. and FORTUNE J. Understanding System Failures, 1984, (Manchester University Press, Manchester)

(16) FORTUNE J. and PETERS G. Learning from Failure - the Systems Approach, 1995, (John Wiley & Sons, Chichester)

(17) GLENDON A.I and McKENNA E. Human Safety and Risk Management, 1995, (Chapman & Hall, London)

(18) KIRWAN B. Human reliability analysis of an offshore emergency blowdown system, Applied Ergonomics, 1987, 18, 23-33

C507/002/95

Emergency planning: a regulator's perspective

A V JONES BSc
Health and Safety Executive, Bootle, UK

SYNOPSIS

As the regulator of health and safety at work the Health and Safety Executive (HSE) has an important role to fulfil in emergency planning. HSE works in partnership with other players acting as a facilitator, using its stock of knowledge and expertise for the common aim of achieving sound arrangements for emergency planning preparedness and response. The paper examines this work and the role HSE has in shaping the legal framework, securing compliance, promoting research and in developing guidance.

1 INTRODUCTION

The theme of this conference is the management of emergencies in the chemical, nuclear, offshore and transportation industries. The Health and Safety Executive (HSE) is involved in all these sectors as the regulator of health and safety and, for sites subject to the Control of Industrial Major Accident Hazards Regulations (CIMAH),[1] the remit extends to environmental protection as well. This is a broad remit but it is important to understand that there are boundaries to our involvement. In the field of work activities HSE is a strong influencer on the emergency planning legal framework, provides guidance and undertakes programmes to ensure compliance. However HSE is not an emergency service and it does not manage directly emergency situations. This can only be done at the local level through the co-ordinated response of industry, local authorities and emergency services. Put simply, HSE's role is to make sure that those who are directly involved through work activities in creating, controlling and mitigating risks discharge their responsibilities adequately.

Why are we considering emergency planning at all? After all there is nothing new about the concept and practice of emergency planning; it has been practised by the emergency services for many years, long before it came to prominence in the industrial field. It was referred to in the 1975 report into the Flixborough disaster.[2] The Advisory Committee on Major Hazards,[3] set up in its wake, emphasised the importance of mitigation (through land use plans and off-site and on-site emergency plans) as one element of a three part strategy for major hazards which also included identification, prevention and control. This strategy was the basis of the Seveso Directive,[4] implemented in the UK by the CIMAH Regulations. It will be retained as the basic architecture of the Control of Major Accidents Involving Dangerous Substances Directive (COMAH)[5] about which more later.

Yet emergency planning is still a live issue. There are several reasons for this:

i) Emergency planning involves the public. It is a paradox that though today we live longer and healthier lives than at any time in history we are more preoccupied with risks to health, safety and the environment than ever before. Recent accidents at Allied Colloids,[6] Hickson and Welch,[7] Associated Octel, Texaco and the Havkong incident at Braefoot Bay[8] have focused public attention on the consequences of major industrial accidents. It can often appear that nothing less than complete success in preventing such accidents is considered acceptable;

ii) It is an inherently complex issue. It involves knowing about the behaviour of industrial plant, substances and that perplexing variable, people. It involves complex interactions between many organisations and individuals. Site management, workers, emergency services, police, health, environment, voluntary agencies and local populations all have to work together in a highly stressful and unusual situation. And on top of all this, during an emergency there is the media's seemingly insatiable thirst for information which companies ignore at their peril;

iii) within most of the sectors which are the subject of this conference there are changes in legislative frameworks which have recently been, or are about to be, implemented. They all address emergency planning. The COMAH Directive will affect on-shore major hazard sites. The Prevention of Fire, Explosion and Emergency Response Regulations 1995[9] were introduced in June with comprehensive duties covering major accidents in the off-shore industry. In the nuclear field the Public Information for Radiation Emergencies Regulations[10] were introduced in 1993 which supplement long established emergency planning procedures contained in nuclear site licences with requirements for information to be made available to the public before and during incidents. In the transport field the UK will have to implement directives which deal with vehicle labelling and placarding which are an important feature of the response to road traffic accidents involving dangerous substances.

In this introduction I have indicated why I think emergency planning is a central issue to all stakeholders in the health and safety system, and particularly to those in the high hazard and high profile chemical, nuclear, offshore and transportation industries. I have, in the most general way, indicated what HSE's emergency planning role is and, importantly, what it is not. The rest of the paper will consider this role in more detail, and along the way I shall point out what I think are the main challenges facing industry and the regulator alike in the future.

2 HSC/E'S ROLE

The notion of HSC/E as a regulator of emergency planning may conjure up images of a body which is concerned solely with enforcement of the law. Providing a legal framework and securing compliance are indeed important functions which the Health and Safety Commission (HSC) and HSE perform but this is only part of the picture. As I indicated earlier, emergency planning involves many different groups who have a direct role in planning for and responding to emergencies at the local level and HSE should also be seen as a partner with these groups in what is essentially a co-operative venture. In fact HSE acts very much as a facilitator, using its influence and its considerable stock of knowledge to bring sometimes disparate groups together and to jointly find a way forward.

HSE plays several roles in emergency planning including:

i) the provision of guidance and advice - HSE has a valuable stone of expertise and knowledge and many contacts with practitioners in this field. It is in an excellent position to assist industry, emergency planners and the emergency services to prepare for emergencies. HSE also produces guidance and other forms of assistance;
ii) the legal framework within which industrial emergency planning operates - there has to be a legal basis for the controls to ensure consistency and to give the public confidence;
iii) the securing of compliance with the law - by undertaking inspection programmes, promoting understanding of what is required etc. to ensure that legal provisions are properly applied and met.

The paper now focuses on these roles.

3 AN EMERGENCY PLANNING FACILITATOR

3.1 Links with organisations and the development of guidance

It has always been HSE's policy to work with stakeholders in order to develop sensible arrangements to guide people in meeting their health and safety duties. At its most formal level HSC operates a network of advisory committees which bring together experts from stakeholder groups such as employers, employees, local authorities, academia and other government departments to advise the Commission on particular topics (such as toxic or dangerous substances) or particular sectors (such as nuclear).

In the field of onshore major hazards HSC is advised by HSE and it's Advisory Committee on Dangerous Substances (ACDS) which has a sub-committee which deals specifically with the major hazards sector. An example of the sort of work which they are involved in, and which is very relevant to this conference, is rewriting HSE's guidance on emergency planning. In 1984 HSE published general guidance on the CIMAH Regulations (booklet HS(R) 21)[(11)] and supplemented this the following year with further guidance on emergency planning (HS(G) 25).[(12)] Although a useful start, much experience has been gained and lessons learned since the mid 80s. There is now a need to update and amplify these documents to include guidance on the new requirements in COMAH. The major hazards sub-committee of ACDS will be heavily involved in this process. To steer the revision of the emergency planning guidance, they have set up working groups which include emergency planning experts from a wide range of backgrounds. The groups also include experts on the environment.[(13)] The ACDS major hazards sub-committee were also involved in the preparation of guidance on preventing risks to the environment from fire water run off which has just been published by HSE.

Earlier work undertaken by ACDS[(14)] included a study of major hazard aspects of the transport of dangerous substances which recommended improved emergency planning for transportation accidents. The report identified four problems requiring further examination viz:

a) definition and clarification of the incident response roles of the emergency scenarios, local and health authorities;
b) how to encourage local authorities remote from fixed hazardous installations to provide resources and to plan for transport major hazard events which might happen;
c) the provision of assistance from one authority to another; and
d) the need for emergency planning legislation to be strengthened.

Another product of the ACDS work was a study of risks from handling explosives in ports[15] which was published this summer. Some members of the group of experts who carried out that work looked at a number of ports. They concluded that the port emergency plans left scope for improvement and commended recently published HSE guidance[16] on the explosives aspects of port emergency plans as a basis for improving existing plans. These examples demonstrate just how thorough and penetrating the advisory mechanism can be.

HSE also works closely with emergency planning practitioners and maintains close links with professional bodies such as the Emergency Planning Society, the Emergency Planning College, the Society of Industrial Emergency Safety Officers and Chemical Industries Association. At the local level HSE inspectors have close contacts with industry and local authorities where they are a source of advice.

As well as guidance HSE publishes reports of incidents highlighting the lessons learnt so that at other installations where these risks might be present, preventive action can be taken. An example is the report on the Allied Colloids fire in Yorkshire in 1992. You may recall that this major warehouse fire led to environmental impact on the nearby river, caused by fire water run off. The report lists 14 lessons of which 4 are specifically about emergency arrangements viz:- addressing emergency warning; contacting the emergency services; inclusion in off-site emergency plans and immediate actions to prevent a mitigate environmental damage; and the provisions of information on potential toxicity of products of combustion from mixed chemical fires.

Of course, off the peg solutions do not always exist and HSE has an important role as an initiator of research. Some current projects illustrate the range of research commissioned by HSE. They include research into the public's perception of risk around major hazard sites; a review of the methods used for alerting the public around sites in the event of an emergency; and research into the response of workers in emergencies under conditions of poor visibility.

The role of international organisations in this field is important too. For example both the UN through its APELL (Awareness and Preparedness for Emergencies at Local Level) programme and OECD through its Chemical Accidents Experts Group are promulgating guidelines and action programmes for prevention, preparedness and response. HSE is an active participant in this work which is influencing programmes and policies around the world, including countries in transition in central and Eastern Europe.

3.2 Communicating risk to the public

I mentioned earlier the idea of partnership and I have indicated the range of organisations which HSE is involved with. I now want to turn to another essential partner in the process, the public. We all have a responsibility to address the legitimate concerns of the public, to be as open as possible, and truthful, and to positively engage them in building trust. It was true some years ago, and it is probably true in some quarters today, that the public have been viewed as passive recipients of information about major hazard activities. Information exchange has been seen as a one off exercise and little attempt has been made to find out what the public's information needs are and whether or not they are being met. I think that one of the major challenges of the years ahead is to engage the public far more in this process so that they are involved as meaningful partners with industry. It is in areas such as this that the law can only go so far. For example the COMAH Directive requires information to be made available to the public. This includes basic details about the site such as its location, ownership, activities and substances which are used or stored. It requires also the provision of information on: the nature of the major accident hazards and their potential effects on the population and the environment; how the public will be warned and be kept informed in the event of a major accident; and what actions the public should take in an emergency.

Thus the directive and existing requirements in the CIMAH Regulations deal with the mechanics of the information exchange but not its effectiveness or how it should be tailored to meet local needs. Effective communication of risk is a complex issue and I can do no more in this talk than indicate some points for you to consider.

The first of these is to do with assumptions about people. There is a commonly held view that the public cannot cope with anything but the briefest and simplest of information. To do more would be to confuse and to frighten. There is also a belief that, faced with an emergency situation, the public will panic first and think later. In fact most research suggests the opposite and successful communication strategies are based on a close dialogue with local communities to find out what information they need rather than assume that the risk creator knows what is best for them.

There is difficulty in defining and measuring successful communication but there are a number of factors which seem to make the process more beneficial. These include the following:

i) senior management support for the communication strategy;
ii) early and continuing involvement of the public (don't wait until the accident happens);
iii) responsiveness to questions and concerns;
iv) open relationships with the media;
v) providing clear and simple explanations of the risks and their implications for the community;
vi) engaging as broad a spectrum of individuals and community groups as possible;
vii) involvement of the community in emergency preparedness activities such as exercises and simulations.

4 THE LEGAL FRAMEWORK

4.1 The approach to regulation of health and safety

HSE has a key role to play in the shaping of the legal framework which regulates health and safety risks arising from work activities. First I will look at the general approach to regulating industrial risk in the UK before looking in more detail at emergency planning.

The shape of modern health and safety law has its origins in the report of the Robens Committee of 1972.[17] Robens identified a number of problems with the existing legal framework: it was too detailed, fragmented and prescriptive. As well as being hard to understand there were important gaps eg. in public protection and in sectors outside of manufacturing. The prescriptive approach had 2 basic weaknesses: it detracted from the principle that it was the risk creator's job to control risk not the regulator's; and it could not keep pace with the rapid technological advance of industry. Robens therefore proposed that the old regime be replaced by legislation which was generally applicable to all work activities and set general health and safety goals. The Health and Safety at Work etc. Act 1974 (HSW Act) ,[18] enshrining these principles, came into force in 1975 and remains the cornerstone of UK health and safety law.

4.2 High hazard sector

Whilst the general framework is set by the HSW Act there are some cases where hazards are high and further specific requirements are necessary for dealing with particular activities. For instance the nuclear, offshore installations, onshore major hazard sites and railways sectors all have legislation which supplements HSW Act; and for onshore major hazards this embraces the environment as well as people. The regimes for those sectors are far from identical but they do share two important characteristics:

i) they require the operator to demonstrate that their activities are safe by either requiring evidence to be provided to HSE or through production of safety cases or, in the case of nuclear sites and sites handling explosives, through adherence to a licence issued by the regulator;
ii) they recognise that incidents could have severe consequences for large numbers of people or the environment; and
iii) they require the production of emergency plans.

4.3 Health and safety management systems

There is one other significant feature of the modern legal framework which I must mention, the importance of safety management systems. The pre-Robens legal framework was based on a perception of safety as a technical problem. This ignored the fact that technical deficiencies are often just symptoms of more fundamental problems rooted in management failures. Hence the HSW Act requires employers to prepare a health and safety policy and to have an organisation and arrangements for putting it into effect. This approach has been taken further in the Management of Health and Safety at Work Regulations and it is a feature of the safety report, safety case and licensing regimes that operators must specify their safety management systems as well as plant technical standards.

Organisations which manage health and safety successfully display a number of common characteristics. The key elements for success are:

i) clear health and safety policies which guide decisions and activities;
ii) organisational structures which create positive safety cultures and secure the involvement and participation of staff at all levels;
iii) planned and systematic approaches to policy implementation using risk assessment to decide priorities and set standards;
iv) systems for measuring performance against pre-determined standards;
v) systems for audit and review which enable the lessons of experience to be fed into revised policies and standards.

It is important to see emergency planning in the context of this total approach to health and safety. Organisations are unlikely to succeed in terms of emergency response unless they have properly addressed all these elements.

They provide the framework within which more specific health and safety processes are undertaken. Those processes start with the recognition of hazards and risk identification.

They are followed by an assessment of likelihood and consequences and the identification, and implementation of systems to prevent or control risk to as low a level as is reasonably practicable. In the context of emergency planning inadequacies in any of these processes could lead to deficiencies in the emergency plan. Moreover mechanisms are required to learn from mistakes and for arrangements to be modified and improved to reduce still further the possibility of emergencies arising. Even the best laid plans may not meet all eventualities. Plans should be simple, be flexible, be capable of meeting a wide range of scenarios and be easily understood by all those who may need to use them.

Apart from the role of employers and operators in preparing on site emergency plans, other key players have responsibilities under health and safety legislation. Local authorities must prepare off site plans and in ports the statutory harbour authority has emergency planning duties. For these plans to work well there should be close working and co-operation between those preparing the on and off site plans and with the agencies who are likely to be involved, - local health, fire and police authorities, voluntary services and other players depending on the circumstances eg. transporters of dangerous goods, British Rail.

4.4 Onshore Major Hazards

To illustrate some of the main features of the regulatory system touching emergency planning I will look now at onshore major hazards. The CIMAH Regulations implementing the Seveso Directive were introduced in the UK in 1984. They established a regulatory regime for the identification, control and mitigation of the risk from onshore major hazard sites. The regulations placed duties on operators to produce on-site emergency plans and on local authorities to produce off-site emergency plans. These duties will be updated in the COMAH Directive which is expected to be implemented in the UK in 1998. The structure and content of COMAH will be broadly similar to CIMAH. It will apply both to people and the environment. There will be two tiers of application of the directive depending on quantities of dangerous substances present on site. Dangerous substances are specifically listed or they are defined by reference to their generic properties e.g. toxicity or flammability. The directive contains an explicit requirement for emergency plans for all top-tier sites i.e. those with the higher quantities of dangerous substances.

The objectives which the COMAH directive sets for emergency plans are as follows:

i) containing and controlling incidents so as to minimise their effects, and limit damage to man, the environment and property;
ii) implementing the measures necessary to protect man and the environment from the effects of major accidents;
iii) communicating the necessary information to the public and to the services or authorities concerned in the area;
iv) providing for the restoration and clean up of the environment following a major accident.

The directive also gives more detail on the data and information which should be included in on and off site emergency plans and these are reproduced in Annexes 1 and 2.

The Directive requires the process of producing emergency plans to include consultation with employees on on-site plans; and with the public on off-site plans; that plans be reviewed, tested and where necessary revised by operators and local authorities at intervals of no longer than 3 years. The review is required also to take into account changes occurring in the establishments or within the emergency services, new technical knowledge and knowledge concerning the response to major accidents.

There are three important features of the approach in the COMAH Directive which are worth noting. First, it is a non-prescriptive approach in the sense that it does not specify precisely what action should be taken to deal with an emergency. It would be inappropriate for regulations to attempt to second guess all the myriad ways in which emergencies could occur and to impose a legal strait-jacket for emergency planning and response. Rather a framework is set in which those with responsibilities can construct plans best suited to particular circumstances.

Second, the approach reflects the need to embrace the key elements of safety management systems which I listed earlier. These elements should be built into the management systems at the installation or site. Operators should set clear objectives for what they are trying to achieve in their plans . They should specify how they are organised to deliver these including the responsibilities of individuals, the resources available to them including training and the way in which co-ordination of activities is to be managed. They should identify hazards and assess risks and from this should flow plans which deal in detail with foreseeable event scenarios. Finally, there is recognition of the importance of audit and review of procedures, the testing of plans and application of the lessons learnt from incidents.

4.5 Other Sectors

The regulatory regime is rarely static. There is a continuing need to keep controls under review to make sure that they remain effective and relevant.

4.5.1 Pipelines

A current example is the proposed new regime for pipelines.[19] The review of pipelines legislation prompted by the complexity, limited range and prescriptive nature of existing legislation was given greater immediacy by the Government's initiative to begin opening up the domestic gas supply market to competition from April next year. It will come as no surprise that the proposed solution is regulations which are goal setting, widely applicable and which embrace the safety management approach.

Emergency planning is an important part of this new package. The approach is similar to COMAH. There will be two tiers of duties depending on the hazards associated with the substance in the pipeline. For the most hazardous substances there are requirements for emergency procedures to be drawn up by the pipeline operator. Local authorities will be required to prepare emergency plans and to keep them up to date. This is similar to the off-site emergency planning requirement in CIMAH. However because pipelines cross local authority boundaries, there will need to be close liaison in producing plans. There are important differences too. For instance, it would generally be impracticable to provide a means of warning the public along the whole length a pipeline. Furthermore, unless the pipeline has been damaged by a third party there may not be anyone in the area to raise the alarm. Also, the requirement to provide information to the public is not appropriate, the chance of an incident occurring at any particular point along a pipeline being so small.

4.5.2 Domestic gas market

The domestic gas market illustrates another, often less spectacular but equally important facet of emergency planning. The opening up of the domestic gas market to competition required a broad legislative overhaul. The move from a single operator in control of the national and local gas distribution network and systems, to a playing field in which there is competition between gas conveyors, gas suppliers and other players in the new market presents some interesting safety challenges. The importance of the effective management of gas safety arrangements will be paramount. The nature of gas emergencies differs significantly from emergencies at fixed sites for example in the chemical industry. The COMAH model is not valid. Two problems have had to be addressed: a) the provision of an emergency response service for gas leaks and emissions of carbon monoxide gas, and b) supply emergencies, that is to say emergencies where demand outstrips supply.

HSC have proposed, in its consultation document,[20] that the gas conveyor should provide an emergency response service for dealing with gas leaks and carbon monoxide poisonings. The gas conveyor's safety case would have to give details which would be scrutinised by HSE as part of the safety case acceptance procedure.

Supply emergencies might be local or national. For example the pipeline infrastructure locally might not be able to cope with demand or there might be insufficient gas to pump into the network nationally. At local level the gas conveyor would be required to plan for emergencies and, as with gas escapes or carbon monoxide poisonings, set out arrangements in the safety case. The risk of a national supply emergency occurring is very low but with a number of gas conveyors on the national network a need is foreseen for a network emergency co-ordinator (NEC). The HSC has proposed that the NEC should be under an obligation to produce a safety case describing arrangements for emergency action in the event of a network supply emergency. The plan would deal with measures to optimise supply, arrangements for safely shutting down part(s) of the network and arrangements for introducing into the network broader specification gas within limits HSC has suggested in its consultative document.

4.5.3 Offshore

Offshore oil and gas platforms have their own special problems and risks are mainly to workers on the platform. Emergency planning needs to take account of the remote location and lack of the usual emergency services available onshore. The primary objective offshore is to provide safe refuges and efficient and effective means of evacuation together with arrangements which will be effective in the recovery, rescue and removal of people to safe locations.

4.5.4 Nuclear

In the nuclear field HSE issues licences for civil nuclear sites. These licences have a range of conditions attached to them including requirements that the licensee makes and implements adequate arrangements for dealing with accidents. In keeping with the rigorous regime applied to nuclear sites HSE is very much involved in scrutinising those arrangements.

5 ENFORCEMENT

So far I have dealt with the legislative framework and the way in which HSE seeks to use its influence and contacts to ensure the framework remains fit for its purpose. I now want to say a few words on HSE's enforcement role.

First let me point out some important features of HSE's approach to inspection. I mentioned earlier the importance of safety management systems. Inspectors make these the focus of their inspections. Inspectors look for evidence that operators have appropriate systems in place and that those systems are being implemented effectively. At major hazard sites a variety of techniques are employed. The CIMAH safety report is a useful tool in this respect. It causes operators to address key areas for achieving safety, to identify the risks which their activities give rise to , the way in which they control those risks and the emergency arrangements required. Inspectors use the safety report, and supplement it with their own knowledge of site conditions, to develop systematic inspection programmes. These increasingly include safety auditing techniques which are highly structured methods for assessing the adequacy of safety management systems. Emergency planning arrangement are scrutinised within these programmes. Over time, the aim is to inspect in sufficient depth and breadth to form reasonable judgements about the quality of health and safety management systems and the effectiveness of their implementation.

When HSE Inspectors assess on site emergency plans they look for evidence that certain key features have been addressed. These include the following:

i) Does the plan contain sufficient detail for the most probable events whilst retaining the flexibility to cope with the largest incident that can reasonably be foreseen?
ii) Are there sufficient resources in terms of personnel and equipment to implement the plan?
iii) Are individual and group responsibilities clearly specified?
iv) Has adequate training been given to those involved in implementing the plan?
v) Are there adequate emergency services both before and during incidents?
vi) Are there adequate arrangements for rehearsing and reviewing the plan?

For off-site plans inspectors will be seeking to ensure that they dovetail with on-site plans. For instance, information about the nature, extent and likely effects off-site of possible major accidents should correspond with that given in the operator's safety report. The arrangements for communications between the site and those involved in the off-site response should be addressed. Emergency information given to the public should be consistent with the on and off-site emergency plans.

As well as assessments made in the course of routine inspection programmes HSE sometimes undertakes co-ordinated national initiatives to pursue particular topics in greater depth. For example, shortly after CIMAH came into force inspectors visited every county emergency planning officer, and regional emergency planning officer in Scotland, to find out how they were approaching their new duties to prepare off-site plans and to offer advice. A more limited exercise recently involved HSE inspectors assessing 100 off-site plans across the country. Nearly a quarter of those plans required improvements and 5 were assessed as seriously deficient. The most common shortcomings reported were a lack of incident specific and site specific detail in plans and a lack of integration with on-site plans. For their part planning officers reported that they often had difficulties in obtaining and interpreting information provided by manufacturers on possible major accidents.

Exercises such as this are a good example of the importance of the link between inspection activities and policy formulation. Inspectors provide valuable intelligence about the standards which are being achieved in practice and this is used to inform future policy developments and for sharper targeting of the inspection process.

6 CONCLUSION

In this paper I have examined a number of ways in which HSE approaches emergency planning. To illustrate essential features of HSE's approach I have looked at fixed on-shore major hazard installations and at how these features are modified when applied to other sectors.

HSE's task as the safety regulator is to shape the legislative requirements on emergency planning, to provide guidance and assistance on their interpretation and application and to ensure that they are met. HSE is not an emergency service. The detailed response to incidents is rightly the preserve of those at the local level, of site operators and the professional emergency services.

Emergency planning is only one aspect of health and safety, albeit an important one. Good management of health and safety lies at the heart of seeing that emergency planning is properly targeted and that it is effective. There will be differences in arrangements between industrial sectors and within the various sectors but whatever arrangements are adopted they should be fit for purpose, deliver good standards and be capable of responding to the unexpected. Mechanisms for review, for learning lessons from past mistakes and failures should be included in those arrangements.

Interest from the public in the risks they face from work activities has never been greater. Their interest at the local level will vary but it should not be ignored. How best to involve them is for local judgement but there is a growing body of evidence that a properly informed and trusting public makes for more effective emergency planning.

In conclusion I will return to one of the characteristics of emergency planning which I mentioned at the outset: the complexity of interactions between individuals and organisations. One of the keys to emergency planning is managing communications between all players so that they have the capacity to plan and act in an appropriate way before, during and after an incident. George Bernard Shaw observed that "the problem with communication is the illusion that it is complete", and so to all those involved in emergency planning, preparedness and response I would say don't be deceived - be prepared for what you foresee and be ready for the unexpected.

REFERENCES

1 Control of Industrial Major Accident Hazard Regulations 1984. Statutory Instrument 1984 No 1902.

2 Report of the Court of Enquiry into the Flixborough Disaster. HMSO 1975 ISBN 011 361075 0.

3 Advisory Committee on Major Hazards 1st Report. HMSO 1976 ISBN 0 11 880884 2
Advisory Committee on Major Hazards 2nd Report. HMSO 1979 ISBN 0 11 883299 9.

4 Commission of the European Communities Directive on the major accident hazards of certain industrial activities No 82/501/EEC (The Seveso Directive) Published in the Official Journal No L 230, 5. 8. 1982.

5 Proposal for a Council Directive on the control of major accident hazards involving dangerous substances (COMAH) Published by The Commission of the European Communities 26/1/94 No COM(94) 4 final; 94/0014 (SYN).

6 HSE. The Fire at Allied Colloids Ltd. A report into HSE's investigation. HSE Books 1993 ISBN 0 7176 0707 0.

7 HSE. The Fire at Hickson and Welch Ltd. A report into HSE's investigation. HSE Books 1994 ISBN 0 7176 0702 X.

8 HSE. Havkong Incident. A joint report by HSE and Marine Accident Investiagtion Branch. HSE Books 1994.

9 The Offshore Installations (Prevention of Fire, Explosion and Emergency Response) Regulations 1995. HMSO. Statutory Instrument 1995 No 743.

10 The Public Information for Radiation Emergencies Regulations 1992. HMSO. Statutory Instrument 1992 No 2997.

11 HSE. A Guide to the Control of Industrial Major Accident Hazards Regulations 1984 HS(R)21(rev). 1990 ISBN 0 11 885579 4.

12 HSE. The Control of Industrial Major Accident Hazard Regulations 1984: Further Guidance on Emergency Plans. HS(G)25. HSE Books 1985.

13 DoE. A Guide to Risk Assessment and Risk Management for Environmental Protection. HMSO. 1995 ISBN 0 11 753091 3.

14 HSC. Major Hazard Aspects of the Transport of Dangerous Substances. HSE Books 1991 ISBN 0 11 885699.

15 HSC. Study of Risks from Handling Explosives in Ports.HSE Books 1975 ISBN 0 7176 0917 0.

16 HSE. Explosive Aspects of Port Emergency Plans. HSE Books 1994. Information Sheet No 3.

17 Safety and Health at Work: Report of the Committee 1970 to 1972. Department of Employment, Committee of Safety and Health at Work. HMSO Command Paper 5034 (Robens Report).

18 The Health and Safety at Work etc Act 1974 Chapter 37.

19 HSC. Proposed Pipelines Safety (PSR) Regulations. HSE Books 1995. Consultative Document CD92.

20 HSC. Proposals for Gas Safety (Management) Regulations. HSE. Books 1995. Consultative Document CD 91.

ANNEX 1

THE COMAH DIRECTIVE: DATA AND INFORMATION TO BE INCLUDED IN ON SITE EMERGENCY PLANS

- Names or positions of persons authorised to set emergency procedures in motion and those who are to take charge of, and co-ordinate, action.

- Name or position of the person with responsibility for liaising with the authority responsible for the external emergency plan .

- For foreseeable events or conditions which could be significant in bringing about a major accident, a description of the action which should be taken to control the conditions or events and to limit their consequences, including a description of the safety equipment and the resources available.

- Arrangements for limiting the risks to persons on site including how warnings are given and the actions persons are expected to take on receipt of a warning.

- Arrangements for providing early warning of the incident to the authority responsible for setting the external emergency plan in motion, the type of information which should be contained in an initial warning and the arrangements for the provision of more detailed information as it becomes available.

- Arrangements for training staff in the duties they will be expected to perform, and where necessary co-ordinating this with off-site emergency services.

ANNEX 2

THE COMAH DIRECTIVE: DATA AND INFORMATION TO BE INCLUDED IN OFF SITE EMERGENCY PLANS

- Names or positions of persons authorised to set emergency plans in motion and those who are to take charge of, and co-ordinate action.
- Arrangements for receiving early warning of incidents, and alert and call out procedures.
- Arrangements for co-ordinating resources necessary to implement the external emergency plan.
- Arrangements for providing assistance with on-site mitigatory action.
- Arrangements for providing the public with specific information relating to an accident and the behaviour they should adopt.

A:\14Sept

C507/003/95

Risk management and emergency planning in the chemical industry

J WHISTON BSc, PhD, MICChem, FRSC
ICI, London, UK

SYNOPSIS

The top priority of the chemical industry is to operate safely and in a manner which minimises risks to human health and the environment. In this it has a record of which it can be justly proud. However, we do recognise that occasional incidents do occur and have in place well planned and exercised emergency response systems to deal promptly with such incidents and contain their potential adverse effects. We also attach the utmost importance to communicating effectively on risks and emergency planning with both employees and local communities where we operate. All these activities are drawn together under "Responsible Care" – the industry's programme for demonstrating continuous improvement in all aspects of health, safety and environmental protection.

INTRODUCTION

The chemical industry occupies a conspicuous and essential position in the economy of the United Kingdom. It is the fourth biggest and one of the nation's most successful manufacturing sectors; its products and techniques are indispensable to all other areas of industrial activity. Our quality of life, and the means to improve and extend this to all, depend upon chemicals and the innovative ability of the chemical industry in healthcare, food production, housing, transportation, communications and, indeed, almost all of the goods necessary to daily life. Chemicals play a crucial role in combatting global problems such as overpopulation, starvation, disease, or environmental pollution.

In Britain we have an important part of the global chemical industry. It contributed £4.6 billion to our balance of payments in 1994 and employs some 280 000 people. It also invests over 7% of its sales proceeds on research and development – nearly £2 billion annually.

The Chemical Industries Association (CIA) was established in 1967 and is the chemical industry's leading trade and employer organisation with a membership of over 200 companies operating from some 700 sites throughout the United Kingdom. It embraces all the industry's trade sectors, product types and business activities.

Indispensable though the industry is, our primary concern is that it should operate safely, recognising, as we must, that it involves activities which are inherently potentially dangerous. A top priority is preventing accidents. Equally, we have to be prepared to ensure that such emergencies as may arise are dealt with to avoid and contain potential adverse affects. The underlying imperative is to have in place well planned and structured management control and response systems.

There are some 450 chemical manufacturing sites in the United Kingdom many of which, as a legacy of the past, are situated in areas of dense population. While there has been increasing diversification of manufacturing operations into low volume speciality products, there is still much capacity involving large quantities of flammable and explosive substances which are subject to processes involving heat and pressure. In Europe well over 100 000 different products are marketed and many of these are hazardous because of their physical and toxicological properties.

The chemical industry has developed considerable expertise and confidence in its ability to evaluate the risks associated with manufacturing and handling chemicals, and to manage their transport, use and disposal safely. Our safety record has shown a year by year improvement since data collection started in the 1970s and, although there was a slight increase in the lost time accident rate per 100 000 hours worked last year, it is now one–third of the level recorded in 1974.

We believe our performance on health, safety and environmental protection is good. But we are not complacent: we are proud to be part of the worldwide chemical industries "Responsible Care" programme and are using this programme to achieve further continuous improvement in this good health, safety and environmental performance. I intend to make some remarks on the overall context of Responsible Care later in my presentation.

HAZARD AND RISK

There are a great many terms used when talking about hazard and risk or the potential for exposure to them. A major problem is that there is no single agreed set of definitions that everyone uses all the time. CIA has recently published an explanatory brochure defining hazard, exposure and risk and explaining what is meant by the assessment, control and management of risk (1).

Hazard is the way in which a product, article or situation may cause harm. Hazards can take many forms and can be thought of as falling into two major groups – constant or intermittent. The hazards from unguarded machinery and from tripping or falling are fairly constant because, if they are present, they can always cause harm. Intermittent hazards, however, are when the product, article or situation that could give rise to harm is not present at one moment but is there the next. This can occur after a change in circumstances which is sometimes dramatic. Examples of this type of hazard include icy roads, severe gales, fires/explosions and sudden escapes of toxic gas.

Exposure is best defined as the extent to which the likely recipient of the harm is exposed to – or can be influenced by – the hazard. Exposure can also be constant or intermittent and must be looked at in relation to the type of hazard involved. With constant hazards, both the presence of potential targets and their distance from the hazard will determine the extent of the risk. When the hazard is intermittent, i.e. when the hazard involves a dramatic change in circumstances, exposure may simply be a question of who or what is present when that change takes place.

Risk is the chance that harm will actually occur. In the same way that risks are part of life, there are risks in all activities that companies carry out. The harm that could occur might affect the company's employees, customers, suppliers, the neighbourhood where the company is situated, the general public, the immediate environment or it may affect the property of the company. However, for harm to occur in practice – in other words for there to actually be a risk – then there must be both the hazard and the exposure to that hazard. Without these being present at the same time there can be no risk.

Much of the work of the chemical industry has been to encourage a proper understanding of risk so that at every stage in the business operation precautions are taken to prevent an incident occurring, appropriate to the inherent hazard involved. For example such analysis covers design and specification of plant and equipment; effects of product exposure; consequence of failure of systems.

Some of the earliest work in the chemical industry was on hazard studies, a technique that is now being successfully used in many other industries. Such studies are a powerful technique for identifying potential problems, but is of course still too late to avoid the hazards. To do this similar studies should be undertaken at the flow sheet stage when decisions are taken regarding product specification, process route and plant location.

All existing plants should have safety audits carried out from time to time. The Chemical Industries Association has produced guidance on the safety auditing process which extends to health and environment as well. There has also recently been increased interest in methods for quantifying the results of such audits using systems such as the International Safety Rating System (ISRS).

Risk Assessment and Control

While the first objective should always be the reduction or elimination of the hazard, there are cases where this is not possible – then the risk must be reduced to a tolerable level. The chemical industry has cooperated closely with the Health and Safety Executive in developing quantified risk assessment (QRA) of manufacturing processes and of the transportation of hazardous products. In particular, we have provided information and assisted in a major appraisal of QRA and of its input into decision making. We believe that QRAs are a valuable tool in helping to understand relative risks, but we have counselled against their use in absolute numerical terms because sound judgement requires flexibility based on experience and rational qualitative assessment of each situation.

Risk assessment of chemical products requires significant time and money and demands good international collaboration if duplication is to be avoided. The initial need is to prioritise, from the many thousands of chemicals in daily use, those for which risk assessment is most urgent. The paucity of adequate information on the health and environmental effects of chemicals demands that scientifically and internationally agreed methodologies must be developed – together with guidelines as to what constitutes acceptable exposure levels and limit values.

The chemical industry worldwide has collaborated with national, governmental and other agencies and has contributed to the development of priority–setting techniques in international fora such as the Organisation for Economic Cooperation and Development (OECD). In the UK, the chemical industry and the government have jointly developed valuable tools in priority setting – many of these have contributed to the development of methodologies adopted by the European Community. Information is being assembled voluntarily and in a way which spreads the considerable cost. Furthermore, the chemical industry has, again jointly with government, developed techniques for the risk assessment of chemicals which recognise the requirement of expert judgement on the part of government agencies to be exercised. There will be a need to explore means by which industry can assist those countries in which technical expertise is less well developed.

As in priority–setting and risk assessment, internationally agreed mechanisms for risk reduction are necessary for those chemicals that cross national boundaries and therefore pose a global threat. However, the provision of hazard information, coupled with safe handling and disposal techniques, and the reduction of exposure are also highly effective approaches to risk reduction. The chemical industry worldwide is participating in OECD and other international programmes for the reduction of risk both in the manufacture and use of chemicals. The results of these programmes will be of significant help in preventative planning for emergencies.

EMERGENCY RESPONSE

Turning now to emergency response, the chemical industry recommends an approach based on assessment of risk, anticipation and preparation. A major publication in this field "Be Prepared for an Emergency" **(2)**, produced by the CIA, was written in conjunction with senior representatives of the emergency services. The basic contention is that all companies should have an emergency plan in place so that there are systems to deal with an incident in the most effective way both in terms of limiting the spread and minimising its consequences. The basic plan has 4 major parts:–

- firstly, assessment of what has happened, for instance a fire or explosion, toxic emission or spillage into a water course. Consideration is given at this stage to hazard and operability studies – the nature of possible events given the plant, processes and substances used. The extent of help and links with the emergency services must be established at this time;

- secondly, a communications plan must be in place for dealing with employees, the emergency services, the media, the neighbouring public and relatives of employees. I shall come back later in more detail to communications, but let me stress that when an emergency takes place it is of paramount importance that there is access by employees, the local community and the media to a senior company professional who can take an authoritative, leadership role;

- a third stage looks at personnel aspects: management available, evacuees, missing persons;

- and fourthly, an assessment must be made of the overall impact of the incident. It is important to know how much of the plant itself is affected and whether the impact of the event extends to the surrounding neighbourhood.

Additionally, companies should undertake regular exercises to test such emergency plans. Various considerations are involved:–

- these exercises must test the comprehensiveness and practicality of the plan. For instance, the allocation of responsibilities between site personnel and members of the emergency services can be established and gaps in communications may be highlighted, particularly through role playing activities;

- initial actions following an emergency may be fairly complex, involving the need to alert a chain of people. It cannot be assumed that such action will take place smoothly without planning, rehearsal and testing of the arrangements;

- we believe that a well practised and communicated emergency plan has the added value of gaining public trust and confidence in a company's emergency capability.

Comprehensive guidance on testing emergency plans is given in CIA's publication "Be Prepared for an Emergency –Training and Exercises" (3).

Sites/The Control of Industrial Major Accident Hazards (CIMAH) Regulations

Much of the advice given could apply to any manufacturing or storage facility. However, much is specific to the production of chemicals. For instance, in drawing up emergency plans some key influencing factors will include:–

- the size of buildings and the nature and quantity of chemicals present;

- the possibility of environmental pollution taking into account the site situation and the immediate surroundings;

- the likely impact on the local community.

The potential for a major emergency is greatest on a site as a result of fire, explosion or large scale release of hazardous substances.

A number of major chemical incidents in the 1970s, including Flixborough (1974) and Seveso (1976), led to the development of an EC Directive on the major accident hazards of certain industrial activities (82/501/EEC). This Directive, adopted in 1982, has been implemented in the UK by the Control of Industrial Major Accident Hazard (CIMAH) Regulations. CIMAH applies to manufacturing sites and storage facilities holding specific substances, or generic categories of substance (e.g. those classified as very toxic, toxic, explosive or highly flammable) above defined threshold quantities. A major requirement is to demonstrate safe operation to the Health & Safety Executive through activities such as major accident hazard identification, arrangements in place to prevent accidents and to limit their consequences, provision of information and equipment, and training.

In addition, sites posing the greatest risk – known as top tier sites – must submit a detailed safety case report to the Health & Safety Executive, prepare an on–site emergency plan and assist local authorities in preparing an off–site emergency plan. Such sites must also provide information to the local public living in the immediate vicinity relating to the nature of the major accident hazards, and on how people will be warned and kept informed in the event of an incident.

Detailed up–to–date knowledge of the site and the immediate locality is necessary to prepare emergency plans. This will include plant layout, the types of chemical processes employed, the products manufactured and, of course, their hazards, together with information on local dwellings, prevailing wind direction etc. The resources that may be required to deal with foreseeable incidents both by site personnel and by emergency services must be considered. The plan should also cover arrangements made to protect people, property, equipment and the environment.

A small site may not have the management expertise and resources to organise extensive self help, and this must be taken into account when preparing the plan. An effective response should be by sharing risks, by delegating functions and, perhaps, by involving assistance form other sites and outside organisations.

CHEMSAFE – The Chemical Industry Scheme for Assistance in Freight Emergencies

Distribution incidents are usually less severe in individual impact, but of course result in considerable public concern about the movement of hazardous chemicals. Furthermore the number of unit movements, and the extent of their geographical spread mean that particular considerations apply since the normal on–the–spot backup will not necessarily be present.

An oleum spillage in the north of England in 1972 is the only occasion on which a life lost was due to the direct release of a bulk chemical cargo on UK roads. At that time, the event caused deep concern to the industry. CIA was then only 6 years old and we were anxious to demonstrate that we were able to organise our industry properly in response to events away from our sites.

As a result in 1974 the CIA launched a project entitled the "Chemical Industry Scheme for Assistance in Freight emergencies" known as CHEMSAFE (4). Its principal objective is to provide prompt technical advice and, if necessary, practical assistance to the public emergency services in the event of an incident involving chemicals in transit so that the adverse effects, including public disruption, are minimised.

There is a standard procedure, in which the manufacturer or consigner responds to emergency enquiries by supplying technical and safety information. It includes a mutual aid arrangement whereby participating companies undertake to help one another wherever possible, often involving a company with a site nearest to the event sending out a competent representative to providing advice at the incident scene. In addition, dedicated schemes have been developed to tackle incidents involving highly hazardous substances such as chlorine (Chloraid) and ammonia (Ammonia Aid).

There is also a long stop procedure in which the National Chemical Emergency Centre at Culham, Oxfordshire operates a 24 hour telephone enquiry facility for the emergency services, to provide information that may not otherwise be available. Personnel can be called to the scene of the incident if required, and they can call on technical assistance from chemical company experts local to the incident scene.

There are other features of the CHEMSAFE scheme but the standard and long stop procedures form the heart of the system. Apart from updating, the basic nature of CHEMSAFE has not changed and I believe it has demonstrated its value again and again over the years.

PRODUCT STEWARDSHIP

Product Stewardship is particularly relevant to today's topic since it seeks to address all health, safety and environmental risks throughout all stages of a product's life from invention, through production processes to ultimate use and disposal. As indicated previously, effective management of risks will prevent incidents. The management system that needs to be put in place therefore needs to identify, document and evaluate as far as possible the actual and potential risks, mainly through hazard identification and exposure reduction or elimination.

Under Product Stewardship, the manufacturer or supplier takes a leadership role in demonstrating total commitment to the principles of risk management and implementing them as part of a continuous improvement process. It seeks to build partnerships between a chemical manufacturer and his suppliers, customers, distributors, agents and others involved in the supply chain. Effective communication, together with feedback of relevant hazard and exposure information within these partnerships, will assist in ensuring that the risks are clearly identified, prioritised and joint procedures for their control and/or risk reduction established. Responsibilities are assigned to each management function – research and development, manufacturing, distribution, marketing, technical service etc. – in striving to minimise adverse health, safety and environmental impacts at all stages of a product's life–cycle. Full details are set out in CIA's guidance booklet on Product Stewardship (5).

An effective Product Stewardship programme therefore combines all aspects of a product's health, safety and environmental management taking into account regulatory requirements, societal pressures and existing best practices. It maximises the care a producer can exercise throughout the supply chain irrespective of ownership or organisational responsibility. As such, it adds value to the product by simultaneously increasing its attractiveness, reducing the risk of harm to people and the environment, reducing the potential of marketing a defective product and, in consequence, reducing liabilities.

For example, advice can be given on the management of waste generated through the product life–cycle as follows:-

- adopt processes leading to waste minimisation using the principles of best available techniques (BAT);
- evaluate feasibility of re–cycling or re–using used and unused product or packaging as a means of minimising waste;
- identify all technical problems and relevant hazards likely to be posed by various methods of ultimate disposal of the product and its packaging – this includes consideration of minor components and impurities;
- ensure that labels and safety data sheets give adequate information on waste disposal.
- ensure all waste product under control is handled in accordance with the principles of the "Duty of Care" and advise customers and users to operate accordingly.

The long–term storage of waste on site is undesirable, but absence of appropriate landfill and incineration facilities means that the best practices must be promulgated for such storage. Safe landfill practices must be employed so that hazardous materials do not present an explosion or pollution risk.

COMMUNICATIONS AND COMMUNITY RELATIONS

I hope I have made it clear throughout my presentation that the chemical industry attaches the utmost importance to communicating effectively with both employees and members of local communities in which the industry's operations take place. Considerable energies are being devoted to improving all aspects of the industry's relationships with its community neighbours, and guidance in this area is drawn from the experience of senior site managers with their own employees and local communities.

The first imperative is that the precautionary and preventative procedures, such as I have already mentioned, should be made familiar to all employees and members of local communities alike. Considerable goodwill is developed through such openness and public suspicions are reduced when the extent of hazards and precautions to minimise risks have been explained.

Over the years, the reputation of the industry in the public domain has deteriorated. We believe that this has happened through a combination of a lack of understanding of the industry's operations and its products throughout the community, coupled of course with considerable adverse publicity as and when incidents, some very serious, have occurred.

Openness solicits trust. At community level the industry finds that some of its strongest proponents are those with the greatest knowledge of its activities and the hazards that may be involved. Sound and reliable communications in an actual emergency are of greatest importance and the situation arising can put enormous load on communications capacity and capability.

All communications systems decay with lack of use. This includes the equipment itself that, for instance, may be installed on site for command and control if an incident occurs. It also includes the communicators in knowing their role, being familiar with the procedures, being aware of the internal and external essential contacts, or having sufficient know–how and confidence to think on their feet since no situation will be wholly without surprises. It is obviously essential that all these aspects should be regularly tested and rehearsed under differing conditions so that those involved operate almost instinctively, and are able to deal with the unexpected.

A golden rule is that in the event of an incident the key audiences which include the people in site; the local community (particularly their leaders); the emergency services; and the media should be able to contact a company representative of sufficient authority and stature who can respond and advise in a calm, open and constructive way.

Such audiences have a powerful influence on the reputation of the company. Moreover, the actions that are taken to keep them informed, the accuracy and the frequency of the advice provided, as well as the way that such audiences are treated, may well affect the future licence to operate at that location. It will also affect the company's standing with the local and international public for along time to come, as we have seen in some recent emergencies not only in the UK but for instance in Germany too.

I recommend a number of guidelines, which CIA has incorporated in its communications and community relations guidance. These include:–

- develop a bank of goodwill through effective communication at all times on the company's business activities, as well as on the procedures that it has in place to prevent accidents;
- don't leave it until there is an incident to communicate – you will not be trusted if you do, as there will be inadequate awareness and understanding of your position;
- ensure that there is overt management commitment to be and to be seen to be involved from the top in communication;

- backup the basic channels of advice and information with systems and procedures that achieve dialogue, understanding, and, above all, involvement of the audiences and sharing of the activity. I have in mind:–

  - community liaison panels
  - local discussion groups
  - open days
  - school/college visits to site
  - participation by company people in community activities
  - media liaison meetings

Straight dealing is essential and a rule for strict observance is that the same information as given to the workforce should be used with the outside audiences. If this should not be the case, differences will be spotted very quickly and public suspicions will be aroused about manipulation and cover–up.

Always remember that suspect information may be disregarded. Inefficient communication, for instance with the media, may well result in material being obtained from other sources which may have established credibility with the public but whose motives are not sympathetic with the company involved.

It is important routinely, and especially if an incident occurs where media coverage is, for example, on local radio, that details are monitored so that false or misleading stories can be counteracted immediately. In particular, it is essential to deal with uninformed speculation such as the number of injuries that may have been caused, or the nature of the hazards to which the public could be subjected as a result of the emergency.

In the event of an incident involving injuries, or perhaps fatalities, it is of course essential to remember the utmost importance of ensuring that the next of kin are informed without delay whatsoever. It is therefore essential to have available, and to make known, a dedicated telephone number for family enquiries to keep this vital function separate from the other communications traffic. Priority and speed on this front is also essential to ensure that the general information broadcast allays the fears of others as rapidly as possible.

CIA has issued a number of comprehensive publications concerning effective media relations, good community relations practice and, recently, a publication entitled "Responsible Neighbours" (6) to ensure that the communications and community relations activity is an essential and integral part of Responsible Care.

RESPONSIBLE CARE

Responsible Care is the chemical industry's commitment to continuous improvement in all aspects of health, safety and environmental (HS&E) performance. It is a voluntary programme of action which demonstrates that sustained HS&E improvement is actually being achieved and, through this, aims to improve the public's poor perception of the chemical industry.

Responsible Care originated in Canada in the early 1980s, was then adopted in the USA, and CIA led its development throughout Europe where the European Chemical Industry Council (CEFIC) now plays a major coordinating role. It is now adopted throughout the world in around 35 countries in three different continents. The UK programme was introduced in 1989 and became a condition of membership of CIA in 1992.

The cornerstone of commitment to Responsible Care is an undertaking, signed by each member company Chief Executive, to a set of Guiding Principles. The signatories pledge that their companies will make HS&E performance improvement an integral part of their overall business policy and that all employees and company contractors will be made aware of this.

Performance improvement is accomplished through voluntary initiatives, self–assessment, training, active community involvement and the development/sharing of best practices from individual companies. The established CIA activity of publishing a series, now approaching 100, of codes of practice (what to do) and guidance notes (how to do it) covering all operational functions and relevant disciplines provides a central focus for Responsible Care.

Responsible Care is, of course, wider in scope than today's subject although accident and incident prevention is a core objective. The programme has a number of elements starting with the guiding principles, and includes indicators of performance, integrated HS&E management systems and communications with employees and members of local communities. One unique aspect of the UK programme is experience sharing through a network of Responsible Care cells where groups of senior site managers meet on a regular basis in 33 regions around the country. Discussions concentrate on the sharing of experience with particular emphasis on risk management and accident/incident prevention.

Responsible Care has received widespread recognition from UK ministers, government departments and regulatory authorities. It has also been acknowledged by pressure groups as helping the chemical industry face up to its environmental challenge. In our view, therefore, it is the leading example of a successful voluntary agreement in manufacturing industry.

CONCLUSIONS

Mr Chairman, I have described in outline the special position of the chemical industry in relation to the theme of this conference and I hope you will agree that we take the safety of our operations very seriously indeed, and have done so for many years. We do care.

I believe that as members of modern society we would all agree that we have to live with hazards and risks whether in personal transportation; the home; sport; or the operations of industry. While risk of harm cannot be totally eliminated - and I do not believe that anyone in society pretends that this could be the case - it is clear that proper precautions can minimise risk and that an enormous amount can be done to prevent accidents and emergencies arising as a result of our operations.

Through setting up highly responsible management systems and procedures; through taking precautions; by systematic planning; and by testing these arrangements, I believe that actual performance can be improved continually. That is certainly the spirit and commitment of Responsible Care.

So far I have made no particular mention of training. However, reference to the various procedures and publications that I have mentioned will demonstrate the underlying importance of effective training for all employees up to the most senior levels in member companies. It is only by thorough training on the disciplines that, in practice, the aim of preventing accidents will be achieved together with effective response should incidents occur. The right course of action to take has to become second nature. CIA, as the industry's official training organisation, has developed a broad curriculum of training and is itself able to organise direct training opportunities. We are also a lead body as regards the development of National Vocational Qualifications covering occupational requirements throughout the industry. Training is, of course, something that must be renewed from time to time, whether this is undertaken on the job or is supplemented by dedicated classroom training experience.

By building on the initiatives that I have described, it is the intention to make the chemical industry increasingly safe in its operations whether on-site, or when its products are being delivered to, or used by, customers. Perhaps our greatest challenge is to achieve a climate of openness, trust and dialogue since, through this, we believe that our performance can be improved still further.

REFERENCES

1. Risk – its assessment, control & management, 1995, Ref PA34;

2. Be prepared for an Emergency – guidance for emergency planning, 1991, Ref RC31;

3. Be prepared for an Emergency – training and exercises, 1992, Ref RC47;

4. CHEMSAFE – the chemical industry scheme for assistance in freight emergencies, 1991 edition, Ref RC32;

5. Product Stewardship – the Responsible Care of products through all stages of their life cycle, 1992, Ref RC46;

6. Responsible Neighbours, 1993, Ref RC68.

All the above publications can be purchased from CIA Publications, Kings Buildings, Smith Square, London SW1P 3JJ. Tel No: 0171 834 3399; Fax: 0171 834 4469.

September 1995

C507/004/95

Emergency planning from an aviation perspective

A OSBORNE BSc H&S Eng, FIOSH, and **D TOMLINSON**
BAA plc, UK

SYNOPSIS

In considering the most appropriate strategy for managing response to any real or potential emergency, there can be no doubt that effective planning and preparation is fundamental, not only to show that the organisational structure is sound and well managed, but that also high standards of competence are in place at every level. The issues facing airports are no different to that facing any major organisation, where the protection of life and property is paramount.

By the very nature of their geography and general layout, major incidents at airports will always tend to be very complex and high profile events calling for timely and integrated response, in conjunction with the emergency services and other airport users. Senior management must ensure that at all times, personnel have received the necessary training and awareness in order to enable a swift and positive response to any incident. In practical terms, this means for BAA constantly testing and validating our plans through a process of ongoing review and monitoring supported by a well managed contingency planning audit process.

1 INTRODUCTION

We in BAA took the view some time ago that our planning for response to any potential contingency or emergency situation had to be capable of withstanding serious challenge in terms of its effectiveness and standards of competence in respect of staff who may be called upon to act at any time. We realised that not only should the plan take into account all potential risks to every phase of airport operations, so far as was reasonably possible, it also needed to address the potential for non-aviation related incidents that might occur either on or off our airports. It was important therefore that from a corporate point of view, we should also ensure that ownership of the plan and in particular, its implementation, was in the hands of those in a day to day sense who were most likely to be called upon to act. However, this does not mean that we were prepared to allow the Organisation to devolve its Corporate responsible to act during a crises. It was recognised from the outset, that responsibility ultimately rests with those at the top of the Organisation.

In other words, those at the sharp end on our airports who are best able to identify the high priority risks have been asked to ensure that our planning is focused very much on real as well as perceived incidents. This also does not mean that we are merely planning for what we assume will never happen, rather, we are engaged in a process of constant testing, monitoring, review, development and improvement in our whole approach to emergency planning and management response.

It is our view that this is a strategy that will pay dividends although we do not want to create an impression that we are in any sense complacent. At each of our airports, a process is in place, to address specific contingency planning issues. Designated personnel at every one of our airports are charged with a responsibility to develop contingency or emergency plans to support a BAA Group wide approach. The review process entails a rigorous and ongoing examination of every facet of an airport contingency plan. However, far reaching this approach may appear at first glance, we felt that even more could be done. We took the view that it was not enough just to have in place a series of plans, we were anxious to ensure that the whole process should be subjected to a thorough contingency planning audit incorporating feedback and remedial action where appropriate.

2 THE AUDIT PROCESS

Some organisations have learned through painful experience, that 'Duty of Care' is a very real commitment for both the organisation and its managers. In effect, it means that a clear and well defined framework policy for emergency planning and management preparedness must be in place, based on a test of 'reasonableness'. The primary objective of the BAA contingency planning audit, is to review current plans in order to examine the effectiveness of those plans, procedures and ability to respond in accordance with requirements placed on the airport both by statute and Duty of Care.

It is inevitable that there will be firm evidence of a commitment to contingency planning although the audit aims to go further in testing the level of understanding and strategic thinking for both BAA and non-BAA, staff at every level. There is of course always a danger that despite strenuous efforts on the part of airport senior management to achieve good communication links, communication may break down; in a way that could be extremely damaging if it occurred during a real incident. Therefore communication is a high priority in all aspects of planning.

The nature and size of an airport together with its somewhat unusual geography, can create a potentially vulnerable situation in the event of a major incident where additional resources may need to be brought in at short notice. Therefore, the audit process incorporates a review process to consider availability of equipment, timescale for arrival on site and scope for securing additional resources if required. The audit team will also be looking for evidence to confirm that management and organisational checklists have been prepared to reflect initial action that needs to be taken in response to an incident in the first half hour or so.

Care should be taken to ensure that responsibility for preparing the appropriate action checklists has not been delegated to inexperienced junior staff who may not necessarily have the range of experience and understanding of how to address high profile incidents To do so, could result that a less than satisfactory process with the shortcomings only being discovered only when something happens.

3 IMMEDIATE RESPONSE

We are constantly reinforcing the message that protection of life and property must always be the priority. At the same time, however, we can not afford to lose sight of the need to ensure compliance with statutory requirements, national guidelines and directives combined with the need to focus on the protection of company assets and operation. The range of incidents calling for an immediate response by key personnel will always be extremely wide ranging and events in recent years have tended to focus on a number of the more obvious areas. These include evacuation, particularly, vulnerable persons such as the elderly and disabled, terrorist acts, bomb threats, fire, road accidents, and chemical spillage, the list is endless. Add to that a need to consider the provision of secondary evacuation points, adequate shelter, good communication and professional advice and one starts to gain a better understanding of the sheer size and complexity facing airport managers when considering their emergency plans.

4 POLICY, STRATEGY AND OBJECTIVES

No emergency or contingency planning process, however well structured, will have credibility unless it is built on a firm foundation comprising clear policy, positive strategy and well defined objectives. Any effective emergency plan above all else calls for consistently sound leadership and management skills in order to ensure proper delegation of tasks. Not only must areas of risk be identified and prioritised, roles and responsibilities must be properly defined, and most important of all, accountability should be clear and unambiguous in the event of any challenge at a later date.

Once the policy has been set by the airport top team, designated personnel should be nominated to undertake the preparation of plans that take into account any potential incident that could reasonably be expected to occur. The plan should also address the process of implementation, showing who is responsible and at what stage. Moreover, sufficient thought must be been given to the level of resources that may be required not just for a few hours, but possibly over a period of days. Ultimately, the plan must take into account the level of skills and training required at each level.

A significant part of the contingency planning audit process concentrates on testing the way in which policy has been set. It will look for example, at what risk analysis and scenario planning has been undertaken, and how effective the plans are that have been developed?

Is there a group responsible for implementation of the plan including the organisation of resources, dissemination of written procedures and the introduction of a comprehensive training programme. Who is responsible for testing the plans, what is the nature and extent of any review to improve on any shortcoming that may be identified during a real or simulated incident. Is the review and debriefing process effective and is there evidence to show that results are fed back to the top management team for action at the earliest opportunity?

There are a number of factors that we see as critical to the success of any contingency plan, these factors include not only the need for clear policy, objectives and strategy, but also a process to ensure that senior management are both aware and in control of the complete contingency planning process. The structure has to reflect very clearly roles and responsibilities for managers and staff together with evidence of links with other agencies and organisations - notably the emergency services. However, as has already been emphasised, the plan is of little use unless at the same time action has been taken to identify the appropriate level of resources and facilities that might be required in response to an incident.

5 MANAGEMENT RESPONSIBILITIES

We know from experience following a number of major incidents, that inevitably, many questions will be asked of managers and staff about how the incident was handled and whether things could have been dealt with in a more professional way The effect of this approach is to call into question the quality and competence of both those dealing with the incident and the organisational structure itself. Therefore, BAA has concentrated a great amount of effort on defining management responsibilities at each level in the organisation. Any organisation will want to send out a very clear message that emergency planning will have the highest priority at all times in the strategic and business planning process.

We all recognise that issues of emergency planning go far wider than a mere statement of intent but how do senior managers satisfy themselves that the policy is understood and applied at every level? Is there a process for Senior Managers to receive regular update reports on the results of monitoring, testing and validation of the plan? Is there also a process to ensure that any new regulation or notes of guidance from whatever source have been incorporated within local emergency plans? To what extent are examples of 'best practice' disseminated throughout the organisational structure and what financial provision has been made for emergency or contingency planning? Some might say that the need for a clear and precise definition of management responsibilities at every level of the organisation is unnecessary perhaps even mundane.

How often have we all heard of examples where a plan has failed because something as fundamental as the call-out list was out of date, the procedures had not been amended or difficulty was experienced in locating facilities such as gas, water or electricity in order to close down part of a key function. We might say that it can't happen to us but the harsh reality for many organisations is that it can and does happen, resulting in allegations and at times claims of negligence together with resulting embarrassment and damaging publicity that follows. Any failure to concentrate on defining clear roles and responsibilities for all who may be called upon to implement the plan is almost certainly inviting at least misunderstanding and confusion and more likely, claims of incompetence at some stage.

6 AUDIT CONSIDERATIONS

We have made much of the importance that we place on the contingency planning audit process and we make no apology for emphasising just how important the audit process is to BAA in its effort to ensure that our plans and procedures are of the highest quality. Therefore an audit team will be looking consistently to identify what local controls are in place to ensure that contingency planning response is maintained at the highest level. We will also be looking to see who is responsible to manage the process and to be reassured that where recent exercise scenarios have identified limitations on the effectiveness of those local controls, action is being taken to address the issue. Key areas such as training, line management responsibilities, monitoring effectiveness and regular reviews of the process should have a high priority. The audit team would also want to be satisfied that resources are matched to need in the event of an incident.

In our view, it is much better if proper thought has been given before hand to these areas, as part of the planning process, rather than to face added pressures at the time of a real incident. We recognise that we must constantly scrutinise and review all elements of the process in order to ensure that there is no opportunity for any breakdown in communication.

7 THE MEDIA

The way in which the public perceive an organisation's response to an incident will be influenced to a very considerable degree by the way in which the media decide to report the incident. It would be wrong to see the media as a threat to good management of an incident. It makes much more sense to harness the enormous communication potential that the media has in order to offer reassurance and advise the public at large that everything is under control and that all reasonable steps have been taken to minimise death and injury. At the first press conference to be held following an incident, the organisation will wish to establish that it is both responsible and competent, and that it is controlling the incident. Statements made at the initial stage can very often go a long way to set minds at rest, and to counter dangerous rumours. However, at the same time, we must recognise that the media will inevitably ask a large number of what may be seen as confrontational and challenging questions.

For example, we can expect to be asked when normality be restored, what is senior management's reaction, what action had been taken to prevent such an incident occurring? What is the organisation's safety record in relation to incidents of this nature? We will also need to have to hand accurate and reliable details of the location of the site where the incident occurred including full particulars of what happened and when. The media will also ask for information about the number of personnel on site at the time of the incident and whether there are any fatalities or injuries. We can also expect to be asked if any of our employees have been injured or killed and, if so, what is their condition and what is being done for both them and their relatives. Has everybody been accounted for? These might seem obvious questions but an organisation that comes across as vague and unsure will almost certainly present an image of incompetence if it does not have the answers readily available.

Even media response is not excluded from the contingency planning audit process. In fact, the aim is to ensure that an effective media handling policy is in place. We would examine levels of training for staff who may be called upon to handle a press conference. We would also look at the content of pre-prepared statements available for initial response, the establishment of media liaison points and the process for access by accredited media representatives to the site of an incident?

8 CONCLUSIONS

An organisation which has failed to prepare an effective emergency planning and management response to any incident must be prepared to accept the consequences that go with inadequate preparation and response. Experience of handling both real and simulated major incidents offers a unique opportunity to subject plans and procedures to comprehensive testing and validation. However the lessons learned must immediately be applied to improve response and enhance awareness at every level in the organisation. The objective must always be to create maximum awareness of the likely impact of a serious incident on the organisation and to emphasise the need to have in place techniques, procedures and systems for incident management and media handling, that projects the organisation as both competent and caring in its response.

C507/005/95

Offshore operations – an emergency management process map

E F BRANDIE
Chevron UK Limited, Aberdeen, UK

SYNOPSIS

Total Quality Management (TQM) is the process used at Chevron to manage our business. It integrates quality principles into everything we do. It has the power to direct change, align and focus effort, and ensure that the needs and expectations of customers, stakeholders and communities are met.

This paper examines application of TQM principles to the critical areas of emergency preparedness and contingency planning.

The 'FADE' process is used to produce an emergency management map, and a five step creative planning model is suggested for use at the development phase.

1 INTRODUCTION

The last decade saw a succession of lamentable incidents in a wide range of industries - constant reminders that human error and material failure may occur even in the most highly developed technologies.

Major emergencies are sudden and somewhat unpredictable, they are also indiscriminate in who they affect. The offshore industry has been impacted by a number of such major accidental events - Piper Alpha, Ocean Odyssey, the Chinook helicopter disaster, the Ekofisk Bravo well blowout, the capsize of the Alexander Keilland; the latter two incidents both in the Norwegian sector of the North Sea.

Such major events create huge impact and have severe repercussions - they almost immediately hit world headline communications networks, they carry very significant immediate and long-term costs, they shape legislation, they negatively impact business (some of the Company's involved in the offshore-related referenced disasters no longer conduct business in the UK) and most importantly, they oft-times result in fatalities and the resultant human suffering, pain and grief which subsequently ensue.

Lessons learned from such disasters are naturally inputs to the Emergency Management Process Map overviewed within this paper.

In reality, however, and thankfully, it is not often that businesses suffer instant, major disasters. More often, emergencies encountered cause a form of partial disruption which develops into a major crisis if not dealt with promptly.

The majority of businesses are efficiently managed and competitive in their own industrial sector - they quickly go to the wall otherwise! - most employees and managers are at ease in implementing day-to-day tasks against which performance and results are measured. Perhaps the one area of vulnerability is if the totally unexpected happened? Could they cope? Have they the people around who could cope? Are plans, procedures and equipment in place to help them deal with the situation?

Revealingly, a survey of Chief Executives in the United States conducted in 1985 by Stephen B Fink, President of Lexicon Communications(1), found that 89 per cent believed 'a crisis in business as inevitable as death and taxes'. Only 50 per cent however admitted they had a plan for managing one. Into which half would you fit?

The process map outlined in this paper is intended as a practical guide for those who might one day have to deal with the unexpected. It centres on the key strategies of planning, preparation, and training.

This map will not equip you to deal with a major emergency, but the base process described may provide a management structure for adaptation to specific, particular requirements. Therefore, no guaranteed 'pathway to success' in Total Quality Management speak - but a mapping process based on risk management principles aimed at managing crisis and reducing potential losses associated with accidental events.

2 THE OUTLINE EMERGENCY MANAGEMENT PROCESS MAP

An adaptation of the Total Quality Management (TQM), FADE concept, **F**ocus, **A**nalyse, **D**evelop and **E**xecute.

As for any process-mapping exercise we need to consider a range of potential 'inputs', incorporate an analytical phase, and produce a range of 'outputs', deliverables from the back-end of the process.

All forms of risk management must begin with the identification of risk (FOCUS), but cannot simply end there. The very identification of the risk necessitates the consideration of what can be done about it - in some cases this will result in an informed decision to accept the risk or to eliminate it, but in the majority of cases it will be an effort to reduce the probability or severity of the loss that is threatened (ANALYSIS).

The Robens Report(2) produced some 20 years ago emphasised that safety and risk reduction is the responsibility of those who create the risks, a principle which Lord Cullen fully endorsed within the comprehensive Public Inquiry Report into the Piper Alpha Disaster(3).

Perhaps the most significant legislative change impacting the offshore sector over the past five years has been the introduction of the Offshore Installations (Safety Case) Regulations(4) and the supporting sets of broad goal-setting regulations.

In simple terms, an offshore safety case is a written document submitted to the regulatory body (the HSE's Offshore Safety Division (OSD)) by the installation owner or operator, which must demonstrate that there is an effective system of management of health and safety on the installation, with arrangements for independent audit of it; and that all hazards which could cause a major accident have been identified and controls put in place to reduce the risks to people to as low as is reasonably practicable (ALARP)(5).

There are two great benefits of safety cases. First, they are the installation owner/operator's analysis, which means they have ownership of it. Secondly, sitting down to identify major hazards and to work out how to best to control them, and to set it all down in a comprehensive document for someone else to assess, is a salutary discipline, believe me, as one who has undergone this process! There is widespread industry recognition that the work of producing safety cases has strengthened safety management systems in the offshore industry.

When the safety case has been compiled and risk reduction measures put in place, there may be a temptation to assume that the spectrum of risk has been countered and no specific, additional counter-measures are necessary.

The major effort associated with safety case compilation will have related to a combination of reduction of the probability of loss and reduced severity of loss - inevitably the former will have resulted in something less than the elimination of risk with a clear possibility that the loss could still happen, whilst the latter will not have affected the probability that the loss will occur.

Much of the analysis of loss probability will have involved quantitative risk assessment (QRA) techniques based on forecasting how the loss will arise and what property, services or resources will be directly or indirectly affected by it.

Like all forecasts, these will be subject to some possibility of error. Losses simply do not happen to order, and there can be no guarantee that when the occasion arises the counter-measures will be fully effective, or that the loss may not happen in a totally unexpected way, or under conditions which are so abnormal that the loss may still exceed the maximum considered to be credible.

This loss prevention or reduction stage in the risk management cycle needs to be accompanied by a certain measure of pessimism about its effectiveness, and to be backed up by contingency planning to reinforce the measures to be taken, and to act as a further means of limiting the consequences of risk, if the chosen measures should prove inadequate or fail entirely.

Lord Cullen also recognised such a need and recommended that the Safety Case Regulations be buttressed by certain other goal-setting regulations dealing with specific features of offshore safety, to give the regime solidity.

Subsequently, the Offshore Installations (Prevention of Fire and Explosion and Emergency Response) Regulations(6) came into force on 20 June 1995. These regulations further promote a risk-based approach to controlling fire and explosion hazards and to emergency response. Specifically, contingency planning, emergency management preparedness and response plans based on major accidental event scenarios identified within the offshore safety case are required to be in place.

3 ANALYSIS LEADING TO THE DEVELOPMENT OF PLANS

The post-Piper Alpha, Cullen recommended legislation referenced above is primarily concerned with the prevention of major accident events and the associated mitigation and control measures. They do not specifically address the complete range of credible emergency scenarios.

Unexpected circumstances such as the Offshore Installation Manager (OIM) reporting a person unaccounted for on his platform; a platform supply vessel Master reporting on VHF radio that his ship has made contact with a mobile drilling unit and tonnes of bunker fuel are spilling into the sea in a designated environmentally sensitive area; a sudden report that a routine structural survey has revealed a major defect on the platform sub-structure at -80 metre depth.

Almost inevitably, these emergency situations are encountered and subsequently notified during ‘out of hours’ periods, usually at weekends or bank holidays, when the onshore-based response and support capability is most vulnerable from a rapid response capability perspective.

This analytical phase of the process should therefore be conducted in a fairly creative and imaginative fashion. There are a number of quality tools available to facilitate this, such as brainstorming, the 'what if?' process and others, but the essence of the term creativity is to supplement the more traditional mechanical analyses which will have already been deployed, to give free reign to consideration of the total spectrum of potential emergency scenarios, inclusive of the intellectual and emotional sides of the process.

4 **PLANNING** (DEVELOP phase of the 'FADE' process)

The 'free reign' analytical phase of the process referred to above is actually an integral part of a five step creative planning process originally purported as a strategic planning model, but increasingly finding favour for emergency planning purposes also.

The five steps are: Analysis; Creativity; Judgement; Planning; and Action (James F Bandrowski - Strategic Action Associates Inc, USA).

Step 1 - Analysis. Look at the Company and its operating environment in new ways, brainstorm hazards and associated risks, project emergency scenarios, get broad involvement to identify as many credible scenarios as possible.

Step 2 - Creative thinking. Use the analysis and insights to look into the future; consider where, and under what circumstances the emergency scenarios could be encountered, project their potential affects.

Step 3 - Judgement. First shape potential scenarios into strategic concepts, then evaluate them. Narrow the field. Combine common-mode scenarios where feasible until there is an agreed list ready for action.

Step 4 - Planning strategy. Create a priority list of strategic actions including how each one will be implemented, resourced, funded and measured.

Step 5 - Implementation. Communicate the plans and reasoning within the organisation and to external agencies where appropriate, monitor implementation continually and retain some degree of flexibility - you have to be able to change and potentially shift priorities as conditions dictate.

There are two distinct aspects to emergency planning. First there is planning focused on countering the immediate effects of the particular emergency situation; to save life, to bring the situation under control and potentially to an end as swiftly as possible, and to limit damage. The second aspect, namely recovery planning, should also, certainly under the Incident Command System (ICS) strategy, begin at the time the emergency is encountered, but is sustained as it is concerned with the future and with ensuring that normal operations/ business can be resumed with the minimum of delay.

The emergency planning process therefore should produce a co-ordinated flow of action, commencing before the emergency is encountered, continuing through the immediate post-loss stage, and extending into the subsequent recovery period.

The more pre-planning that can be done, the more certainty there is about what should be done when the emergency is realised, also the less time will be wasted. It also enables planning to be done at leisure, allowing for widespread involvement and participation, thereby avoiding forcing decisions on the spot in all the confusion routinely encountered during actual emergencies.

Such pre-planning, however, can never remove the need for decisive command and control on-the-spot decisions, there will inevitably be certain unexpected features which cannot be catered for in the pre-planning process - nevertheless, having the broad outlines of what is to be done mapped in advance will ensure that the most effective action possible can be taken in the period which immediately follows the onset of the particular emergency.

Action at this stage needs thought because certain decisions on the spur of the moment to meet the immediate needs of the situation may limit the freedom of action at a later stage of the event and may render recovery from the effects of the emergency difficult or even impossible.

Pre-planning therefore allows time for training the organisation's personnel thoroughly so that they know what is to be done, what each person is responsible for doing and how to do it, and, equally importantly, when to stop doing it.

One of the important offshoots of such pre-planning, the recognition that disaster is possible, enables these activities to be a regular part of staff training and the necessary skills to be integrated into day-to-day functions.

Increasingly within the offshore industry, emergency control and command training is supplemented with assessment and competency processes. Emergency response teams are assessed under realistic emergency response conditions with subsequent feedback to both team members concerned and their employing companies on perceived strengths and weaknesses.

The Emergency Response Plan will be focused on ensuring that the correct response can be made promptly to counter the emergency situation thereby saving life, safeguarding assets, and limiting damage. It is indeed a plan for response because it is the particular nature of the beast which will largely dictate what has to be done.

Under the second stage recovery planning process, the initiative lies more with the organisation as a whole. It is up to the company to plot its own course back to normal operations as rapidly and efficiently as possible.

This may be stating the obvious, but no Company can properly know fully how it functions and its subsequent 'life-support system' until it has subjected itself to a form of analysis - the five step creative planning model previously referenced is considered a useful tool to develop a recovery plan.

Preparation for business recovery is itself a continuous process, and the plan must be reviewed regularly to take account of operational changes, external factors and organisational modifications.

'Ownership' of the recovery plan is a critical aspect; those who need to have the information contained in the plan may not be those who compiled it or by whom it was expected to be used.

There are no 'guarantees of success' with either emergency response or second stage recovery plans, but undoubtedly both increase the chances of timely control, damage limitation and a positive prognosis for rapid return to normal operations.

'Prepare for the worst' is a sound motto - its adoption may even preclude the worst from materialising, due to the critical review processes which such pre-planning calls for, revealing ways of improving loss control management which will avert the major emergency.

5 **TESTING EVERYTHING** (EXECUTION phase)

Only by enacting the 'what if' scenarios against the planned procedures and checklists will anyone become really familiar with them and know if they work.

Practice is vitally important, and any sceptics who feel that there is insufficient justification for the diversion of time, resources and energy away from the primary purpose of the company in order to prepare for an emergency which may well never happen, may need to be convinced otherwise.

Major hazards legislation applicable offshore requires operators to test the effectiveness of emergency response plans, and it is common for full-scale emergency exercises involving the company, agencies and emergency services to be conducted.

Scaled-down versions of major exercises including 'table-top' exercises can be devised and implemented to test 'intermediate stage' severity scenarios (those scenarios which fall between the two extremes of 'major' and 'trivial', the latter probably controllable by common sense loss control response level only).

Exercises are designed to test both the operational and communications response capabilities.

Our company performance standard calls for one major 'unannounced' emergency exercise annually. With other pressures and priorities on their time, there are few managers and emergency response team members who are able to keep absolutely versed with every aspect of the emergency response plan. Familiarity with, and testing the efficiency of procedures, can only be effectively accomplished by regular rehearsal, however, and intermediate 'tabletop' exercises supplement the annual major exercise.

Each exercise is followed by a detailed de-brief in which all the key players state how things went from their standpoint. The tone at such de-briefs is constructive rather than critical, along the lines of 'another time it might be better if . . .' but, inevitably the outcomes from this process are new, positive lessons which are addressed and on many occasions shared throughout the industry. The range of suggestions may focus on aspects of team composition, call-out arrangements, specialised training, facility improvements or support processes associated with the overall response capability.

The object is to have all the processes tested and in place, hoping they will not be needed, but ready and able to meet a major emergency situation tomorrow.

In this respect, emergency management can be likened to an insurance policy. It is when the reminder hasn't been paid that the fire strikes.

6 SUMMARY INFORMATION 'NUGGETS'

- Identify processes for evaluating and cataloguing the range of potential emergency situations (use creativity and imagination!)
- Devise policies for their prevention, mitigation and control
- Formulate strategies, tactics and plans for dealing with each potential emergency
- Identify who and what will be affected by them - inclusive of general business impact
- Roll-out supportive service plans and effective communications channels to those affected so as to minimise loss and subsequent damage to the organisation's reputation
- Test everything

REFERENCES

(1) REGESTER, M. Crisis Management - How to Turn a Crisis into an Opportunity, 1987, Hutchinson Business

(2) ROBENS, Lord. Chairman. Safety and Health at Work: Report of the Committee of 1970-72 (HC Command Paper 5034), HMSO: London 1972

(3) CULLEN, Hon Lord. The Public Inquiry into the Piper Alpha Disaster, 1990, HMSO: London
(4) HSE. The Offshore Installations (Safety Case) Regulations 1992, SI 1992 No 2885, HMSO: London
(5) HSE. The Tolerability of Risk from Nuclear Power Stations, 1992, HMSO: London
(6) The Offshore Installations (Prevention of Fire and Explosion and Emergency Response) Regulations 1995, SI 1995 No 743, HMSO: London

C507/006/95

The Channel Tunnel: planning for safety – an example to be followed?

A MORTON, J NOULTON, and **R MORRIS**
Eurotunnel, London, UK

Planning for an emergency in the Channel Tunnel began even before we started digging. That sounds as though there is something inherently unsafe about the tunnel, but nothing could be farther from the truth. The early planning; the safety-driven design; the care with which we pick our people; our emergency procedures and our training régimes have one aim in mind: to make the Channel Tunnel the safest transport system in the world!

Emergencies will happen, especially in transport. No amount of hardware, software or training can prevent them entirely. The main purpose of our emergency planning is to try to prevent an emergency from turning into a disaster - for the people involved and for the reputation of the company. There is no point in wishing you had thought about emergency planning when you are surrounded in Folkestone by 600 angry, frustrated and, possibly, frightened customers whose cars have gone to Calais following an evacuation.

Almost every aspect of the design of the tunnel system has been driven by considerations of safety. Doesn't one wish that the same could be said about motorways and ferries? And now that it is open, safety is at the heart of Eurotunnel's procedures.

It is not just one tunnel but three, each of which is 50kms. long. The two 7.6m. diameter running tunnels are uni-directional. The cross-section of the running tunnels has been planned to hold a train upright even if its wheels leave the rails. And all tunnel trains have full automatic train protection.

These three simple design features thus obviate the main cause of severe rail accidents - head-on collision, de-railed trains hitting other objects, and driver error.

There is a continuous platform (or walkway) through each running tunnel, giving access to cross-passages every 375 metres. These cross-passages connect to the central service tunnel, 4.8m. in diameter, which is a means of access to and egress from the rail tunnels, for maintenance purposes or in an emergency.

The service tunnel is the main inlet for the normal ventilation flows into the tunnel system. Huge fans on each coast blow air into the service tunnel, from which it percolates into the rail tunnels via non-return valves in every third cross-passage. This has the incidental benefit that the air pressure in the service tunnel is normally higher than in the running tunnels. Smoke from a train fire could not, therefore, follow evacuating passengers into the service tunnel. Special vehicles are used in the service tunnel for maintenance access and for emergency response.

The tunnel is divided longitudinally into three sections by two undersea cross-over chambers. These enable us to switch trains from one tunnel to another and take a section of tunnel out of service. One sixth of the tunnel is closed every weekday night for maintenance and one third at week ends. Although their main purpose is to allow the rail services to operate even during maintenance periods, the cross-overs would also have an important use if there was an emergency. In normal operation, for aerodynamic and other reasons, the two rail tracks are kept separate in the cross-overs by huge transverse doors.

The tunnel is equipped with a variety of communications systems for routine or emergency use and everything comes together in the main control centre at Folkestone which is much, much more than a signal box. From here, the controllers monitor everything that is going on in the tunnel, not just the movement of trains but all of the electrical and mechanical systems as well (ventilation, cooling, drainage, fire detection, power supplies, and so on). Like of our most tunnel systems, the control centre is duplicated by a similar, standby, control room in Calais, which would take over instantly if the Folkestone centre were out of action.

The minimum headway between trains at present is 3 minutes. This means a theoretical maximum of 20 trains per hour in each direction. Even without the enhancements which will one day permit 25 or even 30 trains per hour each way, the Eurotunnel system is on the way to becoming the busiest railway in the world - an honour currently held by a mineral railway in Wyoming!

Half of the capacity of the tunnel is reserved for Eurotunnel's own transport system, knwn as Le Shuttle. Unlike the ferries, Le Shuttle segregates HGV and tourist traffic. HGV drivers have their own separate routes through both terminals and dedicated rolling stock. HGV shuttles which can carry 28 x 44 tonne lorries area currently departing up to 4 tmes an hour from each side, carrying up to 2000 lorries a day.

The tourist shuttles for cars, coaches, caravans and motor cycles can each carry a maximum of 120 cars and 12 coaches. Customers do not have to stay with thier cars, but each wagon - carrying 10 cars in the double-deck version, is separated from its neighbour by fire doors guaranteeing 30 minute fire resistance. In a fire emergency, the train would normally continue its journey to the surface. The motorists in the incident wagon would, meanwhile, have been moved to the safety of the adjoining vehicle.

All passenger shuttles have two locomotives, one at each end, so that, in the event of one failing, the other will be able to take the shuttle out of the tunnel. The locomotives are among the most powerful in the world, generating 7600 HP (or 5.6 MW). It is possible to split a train - the shuttles can be uncoupled internally - leaving an affected section underground while the two other sections return to the surface in different directions.

One shuttle can push or pull a disabled shuttle. We also have diesel locomotives on standby in each terminal, so we could even recover a train should both national grids fail simultaneously.

The remaining 50% of tunnel capacity is available for through trains - the Eurostar high-speed passenger trains and freight trains. All of the rolling stock using the tunnel (apart from some freight wagons) is purpose-built. There is no other rolling stock in the world which could meet the stringent safety standards of Channel Tunnel trains.

We have demonstrated to the governments in many tests that either by transferring passengers to a train in the other running tunnel or by other means, we can meet one of the requirements of our Concession, that, in the event of an emergency, we can bring passengers to the surface within 90 minutes.

Eurotunnel was the first transport operator to prepare and publish a safety case. It is now standard practice for new railways in the UK. We did so before we opened for commercial service in order to demonstrate the safety of the system. It is a sore point with us that our direct copetitors do not produce safety cases.

The governments had set us a target of ensuring that a user of the tunnel would be at least as safe as a traveller over a similar distance on the domestic rail networks. The safety case suggests that passengers will be between 20 and 50 times safer than on conventional railways. Given that modern railways have shown themselves to be the safest form of transport, we believe that we have reached our goal of developing the safest transport system anywhere.

Getting the design and the technology right was only one part of planning for an emergency: the people issues are equally important. In many, many circumstances, it is human beings who take over from machines when things go wrong. The human profiles required for each post must, therefore, be specified as carefully as possible and tested thoroughly during the recruitment process The people who handle emergencies must be able to react rapidly, to think logically under pressure, and to assimilate a variety of information and act upon it.

Too often we hear that a disaster occurred due to a build-up of individual and, perhaps, unrelated incidents. A machine can cope with individual failures, but it takes a human being to cope with a succession of different failures. Training programmes must be developed with this in mind.

The safety culture of a company is also vitally important in averting emergencies. We believe that it is essential that all incidents should be reported. As people may, though, be reluctant to report their own errors we do not, therefore, as a matter of principle, take any disciplinary action for genuine errors if the person responsible reports it.

For every accident that happens, there are 600 incidents. We need to know about those incidents so that we can develop plans or procedures to obviate them in future.

The final element is the emergency plan itself. In Eurotunnel, we have a three level plan:

1. the general arrangements, which apply not just within the company but to the Emergency Response Organisations (fire, police, ambulance):

2. the principles, which go into detail about the equipment available, passenger handling, and so on; and

3. the detailed job-specific operating instructions.

With Eurotunnel, the emergency planning arrangements are twice the fun because we are bi-national. The Bi-National Emergency Plan or BINEP has to accommodate the different conventions of the public authorities of Britain and France in handling emergencies; it has to handle the tricky problem of jurisdiction or lead nation; and has to overcome the differences of language and geographical separation.

A fire main runs the length of the service tunnel with branches into the running tunnel. On each branch, there are two hose unions - one to fit the hoses of the Kent Fire Brigade, and the other compatible with the equipment of the Sapeurs/Pompiers. It is theoretically possible for a French fireman to come through the tunnel and extinguish a fire at the Folkestone Portal, though if that happened I would guess that the incident had progressed via accident and emergency to disaster!

Putting together the BINEP has been a task as long as the construction of the tunnel. We have been exercising the plan as we have gone along from the earliest days of tunnelling.

Nowadays, we simulate incidents and accidents on a regular basis to test procedures, plans and operators at all levels and to various degrees of complexity. From time to time, we have full-blown bi-national exercises (7 to date) so that all parties can practice together.

The cheapest and most flexible form of testing the plan is via table top exercises. These do, however, have their limitations and we therefore decided several years ago to organise Command Procedure Exercises. Using a formidable array of communications equipment, they give the decision-maker a mass of information of the kind that he would receive in a real emergency. These are extensively used by the military and give a much more realistic and meaningful rehearsal of the plan. So succesful has this approach been that we and the company who organised the scheme won a national training award.

We believe that we were the first non-military organisation to use CPX. Up to 100 players are isolated, except for the computer. They are fed information on the incident by higher controllers and have to respond. Observers provide feedback. Both the UK and French emergency response organisations have found this technique valuable.

One side aspect of an emergency which can be overlooked is crisis communications. Even if your operations staff handle an incident well, the press will still put a negative spin on the story if you allow them to do so. The first minutes of an emergency are very important and it is essential that you set the story line by early, accurate and open briefing of the press.

We have equipped our two exhibition centres as press briefing centres for use in an emergency, and we exercise this part of emergency planning just as thoroughly as any other.

Since we began full operation, we have had comparatively few incidents but one particular event would have had very grave consequences for Eurotunnel had it not been handled correctly.

On 9 December 1994, we were operating the tourist shuttle in a limited, 'overture' service, principally for shareholders. We were awaiting the final operating certificate from the governments which would allow us to begin charging motorists a full fare, and we know that the issue of the certificate was imminent.

During loading of the shuttle, a car caught fire because of faulty wiring. This was not unexpected. The safety case had confirmed our own view that car fires on board a shuttle were likely, and we had planned accordingly.

In the event, the systems, the people and our crisis communications worked well. The governments, after a brief delay to examine the incident, gave us the operating certificate that we had asked for, without qualification, and we began commercial operations on 22 December.

In August, nearly 900,000 people passed through the tunnel in perfect safety thanks to our planning for safety.

C507/007/95

Planning for potential nuclear accidents

C R WILLBY and **C POTTER**
Health and Safety Executive, London,UK

SYNOPSIS

In the UK, plans for dealing with the off-site consequences of a nuclear emergency have been subject to continual review and development since the commissioning of the first commercial nuclear power stations in the early 1960's. Legislation requires each nuclear site to be licensed and conditions attached to the licence ensure that operators have adequate on and off-site emergency arrangements. Planning in detail is restricted to emergency planning zones of up to 3.5 km radius but plans are capable of extension in the unlikely event that a very severe accident was to occur. The short-term countermeasures of sheltering, stable iodine and evacuation all feature in plans for UK commercial nuclear sites. Exercises test on and off-site arrangements at regular intervals.

1. INTRODUCTION

This paper summarises the UK approach to planning for potential nuclear accidents. It begins with a description of the legislative framework, which requires nuclear sites to be licensed. It is via the site licence that the main regulatory control over emergency planning is exercised. Off-site emergency planning is based on the concept of a detailed planning zone around each site which defines an area in which the use of short-term countermeasures is planned for in detail. Emergency plans developed by local authorities, the operator and other government departments make provision for the use of the short-term countermeasures of sheltering, evacuation, the administration of stable iodine and restrictions on food and water supplies. Plans are tested in regular exercises which indicate that the approach adopted is workable, although plans are subject to continuing review and refinement.

2. THE LEGISLATIVE BASIS

2.1 The current regulatory framework

In the United Kingdom the main legislation governing the safety of nuclear installations is the Health and Safety at Work etc Act 1974. Under this, certain clauses of the Nuclear Installations Act 1965 are relevant statutory provisions, and these prescribe, inter alia, that no site may be used for the purpose of installing or operating any civil nuclear installation unless a nuclear site licence has been granted by the Health and Safety Executive (HSE). HM Nuclear Installations Inspectorate (NII) is that part of HSE responsible for administering this licensing function. A typical site licence has over thirty conditions attached to it. One of the conditions deals specifically with emergency arrangements. This condition places the following requirements on the licensee:

1. There must be adequate arrangements for dealing with any accident on the site.
2. The licensee has to submit the arrangements to the HSE for approval.

3. Outside emergency planning organisations must be consulted.
4. The arrangements have to be exercised.
5. Licensee's staff must be trained.

The licensee's arrangements for dealing with the effects of an incident or emergency within the detailed emergency planning zone (including the on-site response) are defined in the licensee's Emergency Plan. It is this document that the NII normally approves as a condition of the site licence to ensure that it isn't altered without the agreement of the Inspectorate. The Emergency Plan is usually supported by an Emergency Handbook which provides the licensee's staff with detailed instructions for implementing the Plan. In general, the licensee will liaise with local authorities and other public bodies to ensure compatibility between the plans of organisations having responsibilities for responding to an incident with offsite effects.

On January 1st 1993 the Public Information for Radiation Emergencies Regulations (PIRER) were introduced. These regulations implement a European Community Directive and require, for those sites where there is a reasonably foreseeable radiation emergency: (i) the prior distribution of information by the site operator about possible radiation emergencies and their effects to members of the public who might be affected and (ii) that local authorities have arrangements to ensure that members of the public actually affected by a radiation emergency receive prompt and appropriate information on the facts of the emergency and intended health protection measures. Prior information to individual households is issued by the operators in a variety of forms including booklets and calendars. In the case of calendars, these are usually reissued annually. The distribution of such information in the area around each site is in a zone whose size is agreed by the HSE.

In the event of an emergency, the primary means of communicating information to the wider public would be by way of the media (local and national radio and television). As part of the UK Radioactive Incident Monitoring Network (RIMNET), arrangements have been made for information on radiation levels and advice to the public to be included in national television-text systems.

2.2 Future Regulations

Legal obligations for nuclear emergency planning extend only so far as the site operator. Participation in the planning for responding to emergencies by other organisations having responsibilities (local councils, emergency services etc) is by voluntary agreement. This arrangement has worked well and no problems have been apparent. However for the future, the revised Basic Safety Standards Directive may require legislation in this area. The HSE is currently considering whether it would be appropriate to introduce regulations similar to those that exist or are in preparation for other major hazardous sites, which require the local authority to make, review and test off-site emergency plans (currently the CIMAH Regulations, being revised under the EU COMAH Directive). Whether equivalent legal requirements for nuclear sites would take the form of stand-alone regulations or regulations eventually made under the forthcoming revision to the Ionising Radiations Regulations, has yet to be decided.

3. THE UK APPROACH TO EMERGENCY PLANNING

It is considered that the safety standards used in the design, construction, operation and maintenance of nuclear installations in the UK reduce the risk of accidents which could have consequences for the general public to very low levels. Nonetheless, prudence requires the preparation of emergency plans. The basic principles for emergency planning in the UK are as follows:-

(a) there should be a defined zone closely surrounding each installation within which arrangements to protect the public should be planned in detail (the Detailed Emergency Planning Zone - DEPZ). The boundary of this zone is defined in relation to the maximum size of any accident which can be reasonably foreseen (the 'reference accident'). In most cases this equates with the information zone associated with PIRER. For the older type of reactors (Magnox), this gives a detailed planning zone size of up to 3.5 km. For more modern stations, design basis accident studies do not give significant doses beyond the site boundary. For these sites, a minimum detailed planning zone radius of 1 km is cautiously assumed.

(b) emergency planning needs, however, to be capable of responding to accidents which, although being extremely unlikely, could have consequences beyond the boundaries of the detailed planning zone. In other words, the detailed plans should be capable of extension to a larger area if necessary. To assist planners in checking the extendibility of their detailed plans, the NII have issued a hypothetical severe accident scenario based on a beyond design basis accident in the Sizewell B pressurised water reactor.

4. MANAGING THE RESPONSE TO A NUCLEAR EMERGENCY

4.1 Development of the Off-Site Centre Concept

Prior to the Three Mile Island (TMI) accident in 1979, decision-making for the protection of the public off-site was primarily site-based. The on-site emergency control centre acted as the focus for decisions on protective countermeasures such as sheltering, the administration of stable iodine, evacuation and food controls. The experience at TMI, however, showed the potential for the site to become overloaded with too many tasks to perform. Not only was the TMI site management trying to deal with a major accident on-site, but was also attempting to cope with the considerable demands placed on it by the world's media and by numerous other off-site agencies.

Following a review of the UK arrangements, the concept of the Off-Site Centre (OSC) was introduced at all major nuclear sites in the early 1980's. OSCs are facilities situated some distance away from the site at which support to the affected site could be provided. Typically OSCs were established 10-30 miles from the site, sufficiently far away to make it extremely unlikely that the operation of the centre would be directly affected by any on-site accident.

The key functions of the OSC are:

- to relieve the Site Emergency Controller of the responsibility for controlling the operator's off-site activities;
- to provide a centre at which all the relevant agencies can receive information;
- to formulate advice to the local emergency services on action necessary to protect the public;
- to provide a forum at which the local interactions of the various agencies can be co-ordinated;
- to provide a media briefing centre;
- to mobilise resources to assist the site management in bringing the plant to a safe state.

At TMI the chief spokesperson on the accident had been initially from the utility, which led to credibility problems with the media. It was really only when the independent Nuclear Regulatory Commission took over the main briefing role that a measure of public confidence was restored. The UK Government review decided that a single authoritative and independent spokesperson - the Government Technical Adviser (GTA) should be based at the OSC. The GTA's main duties would be to:

- provide independent and authoritative technical advice to police and other authorities handling the off-site response;
- at media briefings provide, where necessary, an authoritative response on behalf of the Government;
- ensure that the lead government department is kept fully informed on all matters relating to the emergency.

Because of the importance of understanding radiological terminology and the operator's assessments of the course of the accident and off site consequences, the GTA needs to be a senior technical official and would usually be a Deputy Chief Inspector from the NII. The GTA's role is purely advisory. He has no executive functions in relation to on- and off-site actions. The responsibility for off-site actions to protect the public remains with the relevant executive organisations. The responsibility for restoring the plant to a safe condition continues to rest with the operator and, throughout this period is still subject to NII inspection and regulatory control. Indeed, the NII would seek to place an inspector, normally the Site Inspector, in the operator's on-site Emergency Control Centre. The operator's organisation would initially provide advice to those organisations responsible for taking off-site actions until the GTA took up his responsibilities. The GTA would be provided with a technical team which would include secondees from the NII.

4.2 The Development of the LEC Concept

The UK has not suffered any accidents with off-site radiological implications since the Windscale fire in 1957 and off-site emergency plans developed since that time have not been tested in a real event. However, a large number of emergency exercises have indicated that the arrangements based on the OSC would provide an acceptable basis for the protection of the public.

In 1989, the Association of Chief Police Officers (ACPO), which represents all the senior police officers in the UK, were developing an "all hazards" approach to major disasters. This followed experiences with a number of UK disasters in the late 1980's (eg. the bombing of Pan Am 103 over Lockerbie in Scotland). With this development, a consistent approach would be taken by the police and other local emergency services to any major emergency. It is based on the idea that the decision making centre would be established in the regional or county police HQ in rooms which could be adapted quickly to the needs of an emergency. The police Commander would be based at the HQ, as would representatives of all emergency services and other agencies and organisations involved in the emergency response. The police Commander and senior representatives of other organisations would form a co-ordinating group, chaired by the police, which would meet regularly to formulate policy on dealing with the overall management of the incident.

The existing arrangements for responding to nuclear emergencies, based on the OSC, did not fit in easily with the approach being proposed by ACPO and they recommended that the nuclear site operators should move towards this approach. The largest UK operator, Nuclear Electric (NE), examined their emergency arrangements in the light of the ACPO recommendations and decided in 1990 to modify their off-site response arrangements to fit in with this approach.

4.3 The revised arrangements

The NE proposals revolved around the concept of a Local Emergency Centre (LEC), complemented by a Central Emergency Support Centre (CESC). The prime LEC role is to decide on the actions to be taken off-site to protect the public, to ensure that those actions are implemented effectively, and to ensure that authoritative information and advice on these issues is passed to the public. It would be staffed mainly by locally based personnel, including:

- Police
- Local authorities
- Fire service
- Health authority
- Coastguard
- Water companies
- Agriculture department

All of these could be involved in a non-nuclear local emergency. In addition, the following staff would attend the LEC in a nuclear emergency:

- Operator's team
- The Government Technical Adviser team
- Nuclear Installations Inspectorate
- National Radiological Protection Board (NRPB)

- Senior Government Liaison Representative
- Agriculture department radiological experts

The LEC arrangement allows the police Commander to receive authoritative technical advice directly from the responsible adviser. Initially this would be provided by a representative of the operator but would ultimately be provided by the GTA once he or she had arrived at the LEC and been fully briefed on the situation. The LEC is under the command of the police Commander who would be responsible for calling and chairing the 'co-ordinating group' meetings of all the agencies present.

The Central Emergency Support Centre (CESC) is based at the Nuclear Electric corporate headquarters in Gloucester. Its prime role is to acquire and assess all necessary technical data relating to the radiological hazard. Once operational, the CESC takes over the job of communicating with the off-site monitoring vehicles and would collect and analyse the incoming radiological data from both on and off the site. The centre would be staffed primarily by operator's staff, but both the NII and the National Radiological Protection Board (NRPB) would also attend to provide an independent oversight. The CESC would be the prime outside link with the affected site, passing on relevant information and advice to the decision makers at the LEC.

Since the revised arrangements for Nuclear Electric sites became operational in March 1994, exercises have been held at the majority of the LECs. Analysis of the results of those exercises has shown that the new arrangements are working well. The relatively minor problems identified in these exercises have not shown up any substantive deficiencies in the LEC/CESC concept.

The other major operators in the UK have not followed Nuclear Electric's response to the recommendations of ACPO and the 'all hazards' approach. However, arrangements at each of the OSC's maintained by these operators have been modified to accommodate the ACPO proposals. For example, the police Commander will now attend the OSC in person and will take control of the off-site operations from there. This addresses one of the major ACPO concerns and allows face-to-face interaction between the police Commander and expert advisers at the OSC. In addition, procedures for setting up the OSCs are continually being streamlined to ensure that these centres become fully operational as soon as possible. It is the view of the Nuclear Installations Inspectorate that the arrangements at the NE sites (the LEC/CESC set-up) and at the sites which have retained individual OSCs, all offer an adequate basis for controlling the off-site effects of a nuclear emergency, as has been demonstrated at various exercises.

5. SHORT-TERM COUNTERMEASURES

5.1 Intervention Levels

In the short-term, off-site plans consider the application of sheltering, evacuation and the issuing of stable iodine. In addition, food and water restrictions may also be applied very quickly depending on the nature and severity of the accident. The basic principles for application of such countermeasures are based on advice from the NRPB that their use should be justified and optimised. In other words, no countermeasure should be planned for unless it is expected to achieve greater good than harm and that effort should be made to implement it in such a way as to maximise the net benefit.

The NRPB has specified Emergency Reference Levels (ERLs) which are levels of radiation dose to the public which generally need to be averted to justify introducing a given countermeasure. ERLs are formulated in a two tier system of dose levels (Table 1). The lower levels have been recommended as being levels below which countermeasures should not, in general, be taken because the conventional risks and social disruption resulting from the countermeasure are likely to outweigh the benefits. The upper levels have been recommended as being those at which action should almost certainly be taken. At values between these upper and lower bounds of ERL, implementation of countermeasures is desirable but not essential and must be considered in the light of the situation at the time. Application of ERLs will ensure that risks to the health of individuals are minimised. If ERLs are not exceeded, health effects would be very small and could not subsequently be distinguished from the normal incidence of such effects. ERLs are subject to continuing review to reflect international developments in the understanding of radiation risks.

In developing emergency plans the ERLs, together with predictions of the impact of potential accidents and the likely effectiveness of the countermeasures, are used to define site-specific intervention levels. These intervention levels, expressed in directly measurable quantities, are used to trigger advice on protective actions following an accident. This advice would be given to the police who would carry the final responsibility for instigating the necessary measures, taking into account all the local factors at the time.

Table 1. Emergency reference levels for early countermeasures

Countermeasure	Body organ	Dose equivalent level (mSv) Lower	Upper
Sheltering	Whole body	3	30
	Thyroid, lung, skin	30	300
Evacuation	Whole body	30	300
	Thyroid, lung, skin	300	3000
Stable iodine	Thyroid	30	300

5.2 Sheltering

Sheltering is a countermeasure that would be considered around any nuclear site. It affords protection by interposing the shielding effects of a building's structure and reducing inhalation of radionuclides from the radioactive plume. In the off-site plans for some sites evacuation forms the primary short-term countermeasure (possibly supplemented by stable iodine) but, even for these sites, sheltering is likely to be required for the period before evacuation is begun. The areas surrounding the majority of the nuclear sites in the UK are sparsely populated, with few buildings of high multiple occupancy (apartments or flats) and, in general, residential properties are substantially constructed. Reasonably high protection factors are therefore likely for the majority of the sheltering population. In general, there are no purpose built public shelters. There are potential problems where sites have trailer or caravan parks in the vicinity, but the numbers involved are relatively few. It is likely that tourists caught in the open or people who are camping, would be advised to evacuate the area or to seek shelter in suitable public buildings.

There is no national guidance on the maximum duration of sheltering. However, a UK survey has found that people would, in general, be able to shelter for one or two days without too much difficulty. Most accident scenarios for UK installations would not require sheltering from the radioactive plume for more than a few hours. Farmers with animals to tend, may find it difficult to shelter for more than a short time and, if necessary, special advice to such groups would be provided by officials of MAFF.

5.3 Stable Iodine

In the event of a large accident at a nuclear power station, radioactive iodine would predominate off-site effects. Stable iodine tablets provide specific protection from such effects. The UK has adopted the latest WHO recommendations with regard to doses of stable iodine. This is available in tablet form as potassium iodate, 50 mg iodine equivalent per tablet. The product licence for the tablets is of a limited duration and they are replaced at two yearly intervals. Distribution is dealt with on a local level, taking into account such factors as local demography, geography and resource availability. The responsibility for making arrangements for distribution has been devolved to local health authorities, who work within a framework of advice from the Department of Health. In general, tablets have not been pre-distributed. For a few isolated households within 2 km of the site at Sellafield, the local emergency planning authorities have, however, decided that pre-justification is justified. This reflects a pragmatic solution to the problem of distributing tablets quickly to widely separated households. Stockpiles of tablets, for public distribution, are held at various public buildings and police stations, and by the operators. The tablets would be distributed to individual households that have been, or are likely to be affected by a radioactive plume. Exactly how this is done will vary from site to site but may, for instance, involve the use of police and local authority staff.

5.4 Evacuation

Evacuation clearly provides a more complete protection but this has to be balanced against the risks and disbenefits of its implementation. Policy on siting and controls over increase in population around nuclear sites, ensure that evacuation of any affected 30^0 sector of the detailed emergency planning zone should be possible within about 2 hours. Local authorities' off-site plans identify one or more evacuation reception centres which would be equipped to receive and care for the expected number of evacuees. Special arrangements would be made to advise farmers who need to re-enter evacuated areas to tend animals.

5.5 Food and water restrictions

A release of radioactivity following an incident could contaminate grass, crops, foodstuffs and food sources in both the terrestrial and aquatic environments. The Ministry of Agriculture, Fisheries and Food (MAFF) has extensive powers under the Food and Environment Protection Act 1985 to control the production and supply of contaminated or potentially contaminated foodstuffs. It has detailed national plans for responding to all types of nuclear emergency in England and Wales. These are supplemented by regional plans which provide for farmers or producers whose land is affected to be notified quickly by the police or MAFF officials. Countermeasures may involve restrictions on foodstuffs, including milk and vegetables, and the movement of livestock.

It is extremely unlikely that water supplies would be significantly affected by any reasonably foreseeable nuclear accident. However, the Department of the Environment, in conjunction with the Water Companies and the National Rivers Authority, will carry out tests on water supplies and ensure that alternative sources of supply are provided in the unlikely event that it should be necessary.

As the radiological pathways to the public via food and water are indirect, the associated countermeasures are very unlikely to be needed in the early stages of an emergency.

6. EMERGENCY EXERCISES

The site licence requires all licensee's employees at nuclear installations who could be involved in an emergency to be trained for their tasks and to be involved in regular exercises to ensure appropriate team performance. In addition to these training exercises, NII requires regular demonstration exercises at each site. Such exercises, known as Level 1 exercises, are witnessed by NII, and this is one of the means whereby NII assesses the effectiveness of the arrangements, training and resources of the operators for dealing with emergencies.

Level 1 exercises mainly concentrate on the operators' actions on and off site and may not always be based on a scenario involving an off-site release. Such exercises may involve the emergency services and other external organisations. The extent to which the off-site facility is activated varies according to the needs of the operators or as required by NII. The timing and scenario of the exercise have to be agreed with NII.

In addition to the Level 1 exercises which focus mainly on the operators' arrangements, there are programmes of exercises to rehearse the function of the off-site facilities and the wider central government involvement. These are known as Level 2 and Level 3 exercises respectively.

The aim of the programme of Level 2 exercises is to test the function of each off-site facility once every three years. Each such exercise requires the operator to staff the off-site facility and provides an opportunity for agencies with responsibilities or duties to take part and exercise their function as appropriate. A wide range of government departments and agencies usually take part voluntarily in such exercises.

The Level 3 exercise is a national exercise and, in addition to testing the setting up and operation of the off-site facility, includes the exercising of the various government departments at their headquarters and at the central government briefing centre, and the interactions between the various centres. The exercise may last more than one day and is chosen from the programme of Level 2 exercises once each year.

7. LESSONS LEARNED FROM NUCLEAR ACCIDENTS

The regular testing of emergency plans in the exercise programme described above, leads to continual, but generally minor, improvements in those plans. The accidents at Three Mile Island (TMI) and Chernobyl were of such a severity that the UK Government carried out major reviews of nuclear emergency arrangements following each one. As discussed in Section 4, the accident at TMI led to major changes in the UK with the establishment of OSC's to manage the off-site response, an attached media briefing centre, and the introduction of the Government Technical Adviser. In contrast, the review carried out following Chernobyl led to lesser changes, although still with some significance.

A particular concern of the Chernobyl review was the extendibility of existing plans to cater for very large accidents. The Review confirmed the existence of adequate national contingency plans which, added to those established for each nuclear installation, would provide a response to any nuclear accident. Improvements which have been made since Chernobyl include the following:

- Improved consultation arrangements The emergency planning process was strengthened at local level, for example by establishing expert emergency planning groups of the involved parties local to each site. At national level the Nuclear Emergency Planning Liaison Group (NEPLG) was set up. This group meets twice a year under the chairmanship of the Department of Trade and Industry, as the lead Department for nuclear emergencies in England and Wales. It is a forum for those with interests in off-site nuclear emergency planning and aims to identify, discuss and find solutions to common problems. The NEPLG aims to agree improvements in planning, procedure and organisation which might form a framework of advice to nuclear operators and the emergency services.

- A nation-wide scheme for monitoring radiation The Department of the Environment established the National Response Plan, and within it the RIMNET detection system for monitoring radioactivity from overseas accidents.

- An enhanced programme of off-site exercises. This led to the establishment of the exercise programme discussed in Section 6.

- Improved time of response from national government Arrangements were made for the independent Government Technical Adviser to be appointed and transported to the site more quickly and his advisory team was strengthened. In addition a dedicated liaison representative at the off-site facility provides a link with the lead central government department.

- Improved information for the public The HSE issued a revised version of its public information booklet on nuclear emergency arrangements, and site operators and local authorities co-operated in the distribution of local information leaflets. The introduction of the PIRER regulations (q.v. Section 2.1) resulted from an EC Directive arising from a concern about inadequate provision of information to the public around Chernobyl.

8. CLOSING REMARKS

High standards of safety in the design and construction of nuclear plants in the United Kingdom and strict control of operations provide a very high degree of confidence that accidents which might affect the public will not occur. Emergency planning provides an additional safeguard so that even if there was an accidental release of radioactive material, protection could be provided to the public who might be affected. Nuclear emergency arrangements are evolving continually in response to changing circumstances, improved techniques and lessons learnt from emergency exercises. In recognition of this the plans are robust, flexible and under continuous review. This ensures that any changes necessary can be incorporated readily into the relevant contingency plans and emergency arrangements.

C507/008/95

Integration and co-ordination of services

N A ROWE BSc, MIInfSc, and **T O'CONNOR BA, MPIM**
Kent County Constabulary, Maidstone, UK
A C WROCLAWSKI MIFireEng
Kent Fire Brigade, Maidstone, UK

SYNOPSIS

Major emergencies are placed in a historical perspective. How the development of the ability to mount a rapid, large scale response increased the importance of integration, is described. Recent advances in co-ordination with the aim of achieving this necessary integration are detailed, and practical examples of current procedures are provided.

1 HISTORICAL PERSPECTIVE

Emergencies with major consequences for people are by no means a recent phenomenon. The scale of the consequences is dependant upon the area affected by the emergency, the number of people within that area and their ease of escape from it. Thus natural disasters affecting huge areas, such as earthquakes, volcanic eruptions, floods, droughts and epidemics, have caused massive deaths throughout recorded history. Most of us will be aware of the destruction by Vesuvius of Pompeii and Herculaneum with some 16 000 deaths in 79AD, and of the eruption of Krakatoa in 1883, the Tsunami from which killed 36 000. Fewer will have heard of the earthquake at Shensi, China in 1556 which killed an estimated 830 000. (1)

In contrast the scale of 'man-made' or 'man-influenced' disasters has grown with the development of technology. For example, in the field of transport ships have been lost with all hands ever since they began setting to sea, (over 600 died when the Mary Rose went down in 1545), but for land travel large numbers of people combined with speed were not achieved until the advent of the railways in the 19th century, whilst in the air, large capacity airliners are really a post world war II development, and space travel is still on such a scale that whilst it has seen tragedies, it has yet to suffer a major disaster. Similarly, throughout history developing technology has resulted in an exponential increase in our ability to kill each other, via the ultimate man-made disaster, war. Paradoxically, improved technology has also always been a route to increased safety.

The UK is fortunate in not being prone to large scale natural disasters but there are numerous parallels between Victorian disasters and the recent events with which we are all too familiar. The sinking of the Marchioness on the Thames in 1989 has echoes in the collision of the pleasure cruiser Princess Alice with a cargo ship off Woolwich in 1878 which resulted in 700 deaths. The Hillsborough stadium disaster of 1989, when 95 people died by crushing, can be compared to the Saturday afternoon childrens show June 1881 at the Victoria Hall, Sunderland when, in the rush to collect free toys which were being handed out at a single door to the gallery, 183 children were killed (2). In addition 19th century Britain saw numerous rail crashes, theatre fires and pit disasters, and these three classes are very clear examples of disasters whose incidence and severity have been reduced by a combination of improved technology and legislation.

2 THE NEED FOR CO-ORDINATION

It is also true to say that nowadays the severity of disasters can be reduced by the improved ability of individual organisations to respond, and their ability to integrate/co-ordinate their responses, but these developments are more recent and depend upon a combination of improved technology and increased understanding. There are good reasons for this. For example, whilst in 1870 it was perfectly normal for several hundred people to travel across the countryside at 80mph in a railway train, in the event of a crash the ability to communicate the fact, mobilise a large scale rescue, remove victims to hospital - by horse and cart - and tend to them once there, was extremely limited. Therefore it is not surprising that following the clear-up there was a concentration on inquiries to establish cause, followed by legislation imposing standards, inspection etc. to prevent recurrence. Post-incident enquiries are still the norm today of course.

With improvements in roads, increases in the capacity, speed and reliability of motor vehicles, and developments in radio and telephone communications the ability to mount a large scale, rapid intervention grew. Concomitantly, the need for the various components of this intervention to be integrated and co-ordinated for it to be as effective as possible became increasingly recognised.

It is generally acknowledged that war and preparation for war brings accelerated development, and the mechanisms set up to respond to air-raids on major UK cities during the second world war are excellent early examples of effective co-ordination between emergency services, local authorities, public utilities, health services, central government and the military. (3) Exactly the type of co-ordination which is still sought today.

3 THE ROLE OF PLANNING

So, how has this essential co-ordination developed in recent years? Key requirements are information and communication i.e each organisation needs to know exactly what it is doing, and ensure all the other organisations know what it is doing. Obviously this can be assisted by prior preparation of a plan setting out what one intends to do and how progress will be recorded and communicated. Plans are initially a state of mind, the thought processes gone through by those involved in preparing them can be as important as the plans themselves, but they are also a record for future reference and an embodiment of formal agreement to perform particular activities.

In the past, it was common for organisations to produce plans in isolation with emphasis only on their own response or on the very obvious links. For example in the USA many hospital plans made reference to co-ordination with ambulance services, but not with other organisations. (4)

It is now much more generally recognised that if all aspects of intervention are to be catered for, with no omissions and no duplication, then planning should involve discussions between all involved organisations establishing a very clear understanding of what roles each will fulfill, and that these agreed responsibilities should be recorded in the written plan. In parallel with this development has been the recognition that, in broad terms, the roles performed, or responsibilities undertaken, by any specific organisation are the same for any incident. For example local authorities care for the homeless: whether their homes have been destroyed by a terrorist bomb or industrial explosion, made uninhabitable by floods, or unhealthy by deposits of radioactive dust, the requirement to respond and the method of response will be the same. As a result there is a growing trend towards the production of 'generic' plans which apply in all circumstances, supplemented by specific plans which outline additional measures or activities that apply to some particular identifiable hazard, but which do not override the generic arrangements. This ensures that there is a single, basic body of knowledge with which staff should be familiar. Such considerations are particularly important when one considers that, other than for emergency services staff, involvement in responding to a major disaster is likely to be a very rare event.

As mentioned earlier plans must also cover the co-ordination of the activities identified as responsibilities of individual organisations. It is sometimes held that this is particularly important in the UK where, when a major disaster occurs, no individual is in charge and each organisation commands only its own resources. This is in contrast to arrangements in most of Europe. In France, for example, the Prefet, or his delegated Sous-Prefet, is in charge and has absolute command over the emergency services, local government resources, health service and the military. However it is clear that such an individual, in issuing his commands, relies heavily on the advice of each of the organisations involved and that the quality of that advice will depend on their ability to gather and analyse information on the incident and progress of the intervention, whilst translating the commands into activity on the ground will demand effective inter-agency co-ordination at that level. In reality therefore the requirement for effective co-ordination is equally valid for all countries. Case histories bear out this view.

4 THE CURRENT APPROACH

In the UK in recent years the approach to managing one's response which has become commonly accepted as effective is to organise at strategic, tactical and operational levels with full vertical information transfer between levels. In some forces these levels are referred to as Gold, Silver and Bronze. Integration between services is achieved, whenever common interest or joint activity renders it desirable, by co-locating bases or centres for the three levels of response, or alternatively exchanging liaison officers at these centres, thereby achieving full horizontal inter-service information transfer at the strategic, tactical and operational levels.

The above approach has recently been set out in a paper 'Principles of command and control' prepared jointly by representatives of all the emergency services and the local authorities (5) and endorsed by inclusion within the Home Office's publication 'Dealing with disaster' (6). It is best illustrated by providing practical examples of how the principles are put into effect by the fire service, the police and local authorities, but before that there is a common cause of misconception which merits clarification, and another point which is worthy of re-emphasis.

It must be made clear that the three levels of activity, strategic, tactical and operational should not be interpreted as a command hierarchy. The Olympic medal style gold, silver, bronze nomenclature tends to enhance this misconception, and in an attempt to dispel it, Kent police now refer to 'Gold Support' rather than 'Gold Control'. In reality each level is autonomous and makes appropriate decisions without any need for reference to, or approval from, the others. Officers leading activity at the operational and tactical levels may well be equal or superior in rank to those at the strategic level. The strategic level, which is commonly a considerable distance from the scene, and staffed by officers who should not visit the scene, is in no position to direct, or second-guess, the operational decisions of those at the scene. This does not inhibit very senior officers from adopting a mobile role, and involving themselves directly with any aspect of the management of the incident where their authority is needed. The purpose of identifying and defining the three levels is to ensure that officers can focus their attention appropriately confident in the knowledge that wider or narrower issues are being dealt with elsewhere.

The point which should be re-emphasised is that whilst at each of the levels, particularly operational, there is a tendency to internalise and limit outward communication to requests for resource, successful management of the incident is dependant upon the full, accurate and timely passage of information up and out, so that the potential effects on the wider community of operational and tactical decisions can be predicted and minimised.

5 OPERATIONAL AND TACTICAL CONTROL

Considering a hypothetical major incident, as represented in figure 1, it is at the actual site, within an inner cordon established and maintained by the police, that the fire brigade will perform its vital key duty of the preservation of life through rescue, allied in the case of fire, by containment and extinguishing. The importance of these activities is recognised by the fact that they will be led and directed by the most senior officer available, not infrequently the Chief Officer of the Brigade. During this rescue phase other organisations will support the brigade by the provision of physical resources eg. heavy lifting gear, additional portable lighting, and of information eg. potentially hazardous substances present and the dangers they pose; precise details of building layouts and construction.

At the scene, outside the inner cordon, the police will co-ordinate the activities of all services and organisations, with a principle aim of establishing and maintaining appropriate conditions for free flow of personnel and equipment into and out of the site/scene, and unhindered performance of tasks by the other organisations present. Central to this is the identification in conjunction with the other services, of locations for key sites such as the rendezvous point, holding area, casualty clearing station, ambulance loading point and evacuation assembly point. Another major concern of the police will be the need to determine whether a crime has been committed. To this end they will be taking steps to ensure the preservation of evidence, and also making arrangements for the identity of injured and uninjured survivors to be recorded for subsequent witness statements. This latter information will also be of considerable value for the establishment and operation of the casualty bureau.

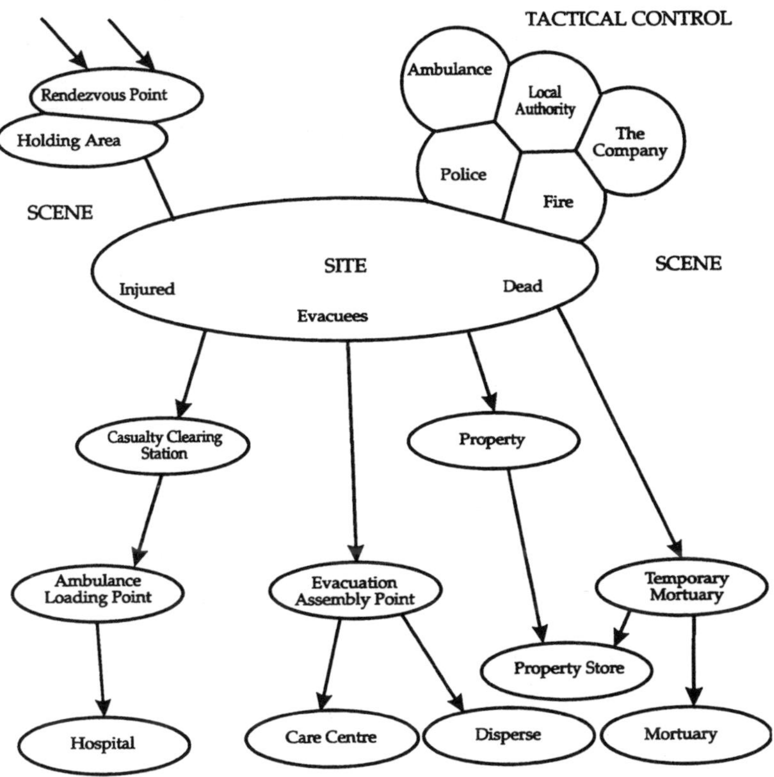

SCENE : The general area of an event

SITE : The specific location of an event

Fig 1 Diagrammatic layout of incident scene

Others involved at the scene may include: health service doctors and ambulance service para-medics tending to trapped casualties at the site and assessing casualties at the casualty clearing station; the company or companies involved eg the airline if it is an air crash, the building owner if it is a terrorist bomb; utilities, if their services have been disrupted or if they are supplementing their services to aid the response; local government eg building control inspectors assessing the safety of structures.

For the emergency services, tactical control is exercised from the forward control points, adjacent to or within the scene. This may of course be some distance from the site. For example, in Kent, for an incident which disables the Kingsferry Bridge, (sole permanent access to the 17,000 population Isle of Sheppey and the busy port of Sheerness), the police tactical control will be established at Sittingbourne police station, some 5 miles from the Bridge, but well within the scene in terms of potential disruption to traffic, and also close to the offices of Swale Borough Council, where the local authority tactical control will be based. Similarly, police tactical control for an incident at Dungeness nuclear power stations will be at Lydd police station, four miles away. In contrast, for an incident at the Channel Tunnel the combined tactical control/liaison point for all key organisations is established at the purpose built Incident Control Centre within the terminal control building. These are three examples of identified potential hazards for which pre-determined arrangements can be made to supplement the generic response. In the more common case of random unpredicted incidents, emergency services may well bring in mobile pods from which to exercise tactical control, or make use of convenient available buildings.

The local authority operational activity will frequently, even at an early stage, be widespread and mostly external to the scene, for example: assistance with traffic management by road coning and signing; establishment and management of rest centres and friends and relatives reception centres; provision of immediate psychological care. Tactical control of these activities will be exercised from the offices of the district council involved, although some of the resource will come from the County Council, and integration with emergency services activities and requirements at the scene will be achieved through exchange of liaison officers.

6 STRATEGIC CONTROL

Strategic control is commonly exercised from the headquarters of the organisations concerned, although once again there are no hard and fast rules, for example, at the Clapham rail crash the police tactical and strategic controls were established on separate floors of a convenient public house adjacent to the cutting, whilst in Cleveland, a compact, heavily industrialised county, it has been predetermined that police tactical and strategic controls will be established at opposite ends of the same corridor in the police headquarters.

Each organisation will have its own strategic issues to consider with regard to resource provision. For example if a fire brigade deploys 30 appliances to a massive fire, how does it, in the relatively short term, provide adequate cover to the rest of its area and, over the slightly longer term, maintain manning levels? Mutual aid from adjacent brigades is only part of the answer. The ambulance service may have similar problems which, in their case, can be partially solved by diversion of non-emergency transport to the British Red Cross and St John Ambulance Brigade. The hard-pressed health service hospitals may have to postpone waiting list operations.

Over a much longer term, local authority social services will have to balance the amount of care they devote to the newsworthy 'glamorous' victims of the disaster against the pressing requirements of their equally deserving normal customer base. All these decisions have to be made by organisations separately, but they may have implications for the others involved, regarding their expectations of service, and there will be other strategic aspects of the response which are far more inter-dependant. For these reasons, whilst co-location of strategic controls is not feasible, full liaison via a strategic co-ordinating group is highly desirable.

In the last two or three years significant steps have been taken towards formalising arrangements for such a group. Two specific examples relate to emergency plans and procedures already described at this conference: the Strategic Support Team (SST) established at Kent Police Headquarters in accordance with the Channel Tunnel Bi National Emergency Plan should a major incident affect the tunnel; the Local Emergency Centre (LEC) approach developed to handle major emergencies involving nuclear power stations, which has resulted in Nuclear Electric funding purpose built and equipped premises at police headquarters whose forces include a nuclear power stations within their areas.

The proposed role of the SST and the LEC is identical, and should the SST be required it would almost certainly be established in the LEC premises. In fact it is now agreed that for any sufficiently large scale multi-agency emergency in Kent involving the police, then a strategic coordinating group with appropriate membership would be set up at the LEC.

For an example I refer back to my previous mention of the Kingsferry Bridge. Should this be severely damaged, causing a closure in excess of a week's duration, then an LEC grouping would certainly be set up. Initial membership would include the emergency services, Kent County Council, Swale Borough Council, Medway Ports, Railtrack the health service and the military (particularly for helicopters). This might shortly be supplemented by commercial helicopter operators, ferry operators, food retailers, Department of Transport and the Home Office (Sheppey has three prisons). Discussions would centre around not only the most effective way of replacing the 26,000 vehicle movements a day which currently service and re-supply the Isle of Sheppey, and transport its population to and from the mainland, but also upon the allocation and prioritisation of whatever transport can be made available in the short term. Joint preparation and agreement of uniform public information announcements and media releases would also be high on the group's agenda.

7 THE LONGER TERM

As the incident progresses and the immediate impact is dealt with, emergency services involvement will decrease. The police may still be maintaining the casualty bureau and pursuing a criminal investigation. If the incident included fire the fire brigade may still be investigating the cause. For the most part however, activity will focus upon the assessment and handling of the longer term consequences of the incident, and the restoration of normality. At an appropriate point during this change of emphasis, the co-ordination role will pass from the police to the local authority. The physical scars may take years to repair, for example the damage caused by the Towyn floods and the City of London bombs. The mental scars may take us even longer, as Kent Social Services staff who dealt with Herald of Free Enterprise victims or Liverpool social services staff who dealt with Hillsborough victims will testify.

At some time during this period, probably within a year, there will be an occasion at which all those involved in the incident are re-united, and all their actions, including the effectiveness of their integration, are examined in detail: the inquiry. Whilst it may still be held that the inquiry's primary purposes are to identify the cause of the incident and make recommendations to prevent recurrence, nowadays the response also is subjected to intense scrutiny, and it is unusual for this aspect, including co-ordination, to completely escape criticism. It is clear that there have been advances in recent years. We should also perhaps be grateful that one of the causes of inefficiency is the rareness of hands on experience due to the infrequency of such major disasters. Further improvements are still desirable however, so there is no room for complacency.

REFERENCES

(1) ALEXANDER, D. Natural disasters 1993, (UCL Press), pp. 594-598

(2) HAMER, M. Lessons from a disastrous past, New Scientist 22/29 December 1990, pp. 72-74.

(3) SANSOM, W. Westminster in war, 1947, (Faber and Faber) pp. 104, 128-131.

(4) AUF DER HEIDE, E. Disaster response, 1989, (C.V. Mosby), pp. 39-40

(5) ACPO et al. Principles of command and control, 1994, (Home Office).

(6) HOME OFFICE. Dealing with disaster, 1994, (HMSO). pp. 12-13, 47-51.

13 September 1995
C:\DATA\PROJECTS\TRAINING\IMECHPA1.WPD
3407

C507/009/95

Education, training, and research in emergency planning and management

J E STRUTT
Cranfield University, UK
J R A LAKEY
John Lakey Associates, Gravesend, UK

1. INTRODUCTION

Over this century, the scale of industrial activity, together with supporting systems and infrastructure, have increased enormously to meet the needs of a growing world population aspiring to material wealth. Many of these developments involve industrial processes which, by their very nature, expose the work force and the public to life threatening hazards. Over the same period the incidence and severity of major accidents appears to have increased and, as indicated in Table 1, they have occurred across a wide range of industrial and transport sectors. As the frequency and severity of major accidents have increased, so has public concern. In the UK and elsewhere this has been reflected by changes in legislation which tends to follow major accidents. Since the 1974 Health and safety at Work Act, UK legislation has increasingly focused on the application of risk management principles, requiring industrial organisations to set acceptable safety targets and to demonstrate, by means of risk analysis, that the targets are being met. For those industries with the potential for major accidents e.g. nuclear, offshore, chemical, aviation rail etc. the demonstration takes the form of a safety case. An important element of a safety case is the management of emergencies.

For industrial processes, emergency planning is typically required wherever there are concentrations of people, either the work force or the public, located close to significant inventories of toxic, flammable or explosive substances. Large, uncontrolled releases of hazardous substances are rare events, however, they must be planned for, as experience has shown that such events are possible even in well run operations and, when they do occur, they can escalate rapidly, leading to multiple fatalities, injuries as well as damage to assets and the environment.

The avoidance of a disaster will depend on the knowledge of the emergency management team and its ability to control events, to prevent escalation and to successfully plan evacuation escape and rescue, under adverse and highly stressful conditions. What are the special qualities needed in a management team if disasters are to be averted? Can the required qualities be taught and what are the methods? The objective of this paper is to outline the education, training and research requirements for emergency planning and management. The paper begins with a description of typical accident sequences, this is followed by an outline of the principle training methods which are currently being used and the impact of human and organisational factors, the effect of stress on human reliability and implications for training and research.

2. ACCIDENT ANALYSIS

The study of accidents, provides a useful input to the training and research requirements of emergency management. Fig.1 shows a simplified event tree which outlines the sequence of events that many major accidents take. The key stages are the initiating event, loss of local control and escalation followed by failure to evacuate or escape.

Initiating Event: The majority of major accidents are triggered by some form of human error or failure of equipment. The trigger event itself may be quite small and relatively insignificant in scale and is often associated with routine plant activities. In many cases there are no warning signs. The underlying causes of such events, however, are complex with a range of interacting human, organisational and hardware factors.

Loss of Control: The ability to control an event at an early stage depends critically on prompt human reaction together with the availability and integrity of emergency equipment and safety control systems {e.g. emergency shutdown systems, fire or gas control systems, water deluge systems, blow down and venting systems etc.}. Such systems must be robust and capable of withstanding, the "loads" imposed by initiating events as experience has shown that the initiating event may prevent some of the emergency control systems from functioning, resulting in a reduced capability to control the incident and a more rapid escalation of events. This has led to a requirement in the offshore industry for studies to be made of the vulnerability of emergency systems to support the installation safety case.

Escalation and Spread: This is the point at which an incident usually escalates from a minor to a major incident. In the case of fires and explosions, for example, the rate of escalation will depend on the scale of the initiating event and on the inventory of materials as well as on the design and construction of the plant and surrounding buildings. Incidents which escalate rapidly increase the potential losses of assets and life, escape routes may become blocked and evacuation of plant personnel and the public may become more difficult.

Failure to Evacuate: The greatest impact on potential loss of life, arises when evacuation of personnel is impaired. This usually makes the difference between those disaster with large numbers of injuries /fatalities and disasters with low loss of life. Evacuation and escape can be greatly facilitated by design and by provision of additional escape routes in buildings, protection of escape routes, temporary refuge areas, and transport systems to provide rapid evacuation etc.. However, in the event of a major escalating incident, experience has shown that human response, leadership and organisation during the incident has an enormous impact on the outcome.

2.1. The risk of a major accident

The risk of a major accident is dependant on the Frequency (F) with which an accident occurs and the potential loss of life N_f. The risk is generally defined as:-

$$Risk = Frequency\ of\ incident \times Potential\ Loss\ of\ life \tag{1}$$

Risks are traditionally represented on a log-log graph (FN plot) for which a straight line with a slope of -1 implies constant risk. Fig.1a is a simplified event tree model of an accident which includes the key stages described above. The four stages can be considered as system states with transition probabilities P_a, P_b, P_c, P_d. The four end points of the tree represent the possible consequence categories.

Fig. 1b is the risk diagram corresponding to the event tree in Fig.1a. This diagram illustrates how the risk changes, from a minor incident to a major accident, as an incident develops. Minor incidents are more likely than major accidents but the risk, when defined as equation (1), may remain more or less constant but with increasing risk uncertainty (implied by the size of the event boxes in Fig. 1b) as events become less and less frequent.

From Fig.1, the risk of a major accident (event 4) is given by:-

$$Risk\text{ (event 4)} = \{F_a.P_b.P_c.P_d\} \bullet N_{f4} \qquad (2)$$

Where, F_a is the frequency of the initiating event (a) (e.g. major leakage of flammable or toxic substance), P_b is the probability of a transition from control to loss of immediate control, P_c is the probability of escalation to a major incident and P_d is the probability of failure to escape. N_{f4} is the potential loss of life associated with end event 4, given the preceding events. Fig. 1 and equation (2) illustrate that the risk of a major event is dependent on the effectiveness of management in reducing (i) the frequency of initiating events and transition probabilities P_b, P_c and P_d and (ii) reducing the potential loss of life.

A loss of containment accident can be made small by increasing the reliability of equipment through improvements in design, construction and maintenance of equipment as well as human reliability improvements through education and training. Likewise, the transition probabilities can be reduced at each phase in the development of an incident by improvements in the reliability of humans and hardware together with increased damage protection strategies such passive fire protection, blast protection etc..

However, given that an initiating event of some magnitude has occurred, the designed-in protection measures alone will not normally be adequate to prevent major losses. The effectiveness of management to intervene, to control and direct events and to limit the consequences will then have a major impact on the final outcome. In effect, management have the task of preventing the transitions from occurring and reducing the injuries /loss of life as an incident progresses

3. EMERGENCY MANAGEMENT PLANNING

In the event of a major accident resulting in the release of toxic or flammable materials into the environment, the effectiveness of measures for the protection of the work force and the public will depend on the advance preparation. An emergency response plan must be developed to control and limit the consequences of the accident. This is now a legal requirement , not only for nuclear plant, but for all process plants which fall within the applicability of the CIMAH regulations, for all offshore installations and more recently for operators of pipelines, both onshore and offshore, transporting toxic or flammable material.

The accident scenarios for planning purposes will range from those with little or no off-site consequence to those having significant consequences off site, even though these accidents may have an extremely low probability of occurrence.

The key objective of an emergency management system are *to prevent or reduce the likelihood of consequential loss in the event of an emergency occurring*. This is achieved by (i) developing emergency plans, (ii) exercising the plans (iii) updating the plan in the light of exercise and real incident experience and (iv) training personnel to manage major emergencies through simulated exercises. As illustrated by Fig.2, these are also elements of a control loop. Fig.2 shows two parallel control loops; an exercise loop (the model) and the real event loop (reality). The common point is the incident control model which represents the perception of what is actually occurring during the real or simulated. The exercise loop provides a simulation of procedures and actions which are expected to occur in a real emergency. At the start of a real incident, the perceptions of reality will be based entirely on the simulation data. The effectiveness of the procedures and the response of ER team is therefore dependent on the degree of correlation between the *model* and *reality*. This reinforces the value of training using simulators. It is arguable that research into the use of Virtual Reality Technology would lead to major improvements in training facilities as well as providing an experimental field for industrial psychologists.

3.1. Components of an emergency plan.

An emergency plan is a document which provides the basis of the procedures to be adopted in the event of an emergency. Emergency plans are based on prior experience of procedures and methods that have been found to work. The main elements of a plan are listed in table 2. These include such items as the type of incident , communication links, technical information, emergency procedures etc.

The plans must be adequate for dealing with an emergency in the event of a major accident with the various components of the plan being tested and checked during development. The plan must therefore be practicable and recognition should be given to the type and magnitude of the events and their associated consequences (risk assessment) for that specific facility. The emergency procedures must include monitoring of the release and mitigation of the consequences by the efforts of emergency response teams (ERT) who are responsible for the application of protective measures and communication within the organisation and with the public.

The Off-site Emergency Plan should be designed to complement the On-site Emergency Procedures so that the point at which procedures might fail is given appropriate attention in training. The education of the ERT should enable them to extend the plan. Training is used to give participants familiarity with their expected response.

3.2. Level of resources

To avoid wasted resources, the detail with which emergency response planning is carried out should be related to the level of risk. The question arises as to the prioritisation of resources and expenditure. Should more emphasis be placed on more frequent less severe incidents or the less frequent major incident ? The usual approach is to prepare a more detailed plan for response close to the site, related to the more likely accident scenarios, with the capability for extension of actions if a less likely, but larger accident, occurs. Although it is not usually practicable to develop a completely detailed response for every conceivable emergency situation, advance planning has the effect of creating a high order of preparedness for any event. It must be recognised however, that major accidents which have the potential for many fatalities can have a catastrophic effect on production and on the viability and credibility of the enterprise and, to many, the only acceptable view is to give highest priority to management of both kinds of accident.

3.3. The emergency management organisation

The USA Nuclear Regulatory Commission (NRC) in association with the Federal Emergency Management Agency (FEMA) have provided useful advice on the task of the management organisation for response to nuclear emergencies [1,2]. The elements of a nuclear emergency response organisation are given in Table 2.

The philosophy and structure is well thought out and applicable to other emergencies. The advice is derived from the experience of the Three Mile Island Reactor Accident. Experience on exercises and with other incidents in the nuclear industry has lead the Institute of Nuclear Power Operations (INPO) to issues guidance on emergency preparedness and to offer a series of courses [3] designed to enhance the performance of emergency response organisations in the utility companies. INPO recognises that the gap between the desired response and the resources available can be high. When the desired response falls in this "gap" the protection given to the public depends on the education and experience of members of the response team especially those who are engineers and scientists. The "gap" is where professional skills (engineering judgement) are demanded, and where precisely defined training schemes and rule driven responses are likely to fail. This problem probably occurs in many other emergency situations and therefore may be a neglected area of the engineer's responsibility for safety of the public. The general principle is widely recognised and the UK Hazards Forum has published [4] a syllabus for an undergraduate awareness course on an engineer's responsibility for safety. It emphasises the duty of care towards the health and safety of the public. A possible response to the INPO "gap" would be the extension of this syllabus to cover the engineer's rôle in the response to emergencies.

3.4. Organisational integrity

Few organisations provide facilities and services exclusively for emergencies but all should consider them as an integral part of operations in their organisation. The interfaces between organisations are the weak part of most plans. Kenney [5] recommends that organisations should use their day to day control systems (the word control being used to include both procedures and equipment) as a foundation for their emergency plans and procedures. He remarks that "bolted-on" systems suffer from bolt failures (either from metal fatigue or loss of tensile strength)! He also stresses the importance of auditing the emergency capability. He recommends that audits should verify the systems for identifying that an emergency plan is required, for developing the plans, for interfacing the plans with outside agencies, for criticising plans and ensuring changes are implemented and for assuring that plans meet all required legislation. Drabek [6,7] also adopts a "stress-strain" perspective in the study of emergency organisational structures which he finds to be characterised by inconsistency, fragmentation and conflict. He identifies skilled managers as those who have become skilled in mapping such strain patterns and in identifying coping strategies that are appropriate for the community in which their organisation is nested.

3.5. Communications with the Public

There is an increasing requirement both in legal and moral terms for industry to inform the population of safety and health hazards to which they are exposed. For this reason, communication with the public and the media is widely recognised as a key part of emergency planning. Prior to and early in an incident, communications will have two principle goals; firstly to reassure the local population and secondly to provide information about protective actions that should be taken in the event that an emergency should affect them. These are in fact conflicting goals and illustrates one of the difficulties of communications[25]. Public messages are subject to social amplification [26] distortion and feedback resulting in unexpected reactions. As testified by Shells experience related to the disposal of the Brent Spar, companies large and small ignore social amplification at their peril.

The Chernobyl Reactor accident demonstrated that, in a situation with environmental contamination, the need for information will be enormous and that preplanned resources will need to be available centrally as well as locally to cope with the situation. Considerable thought must be given to the ways in which the public should be informed about the actual and predicted problems, e.g. levels of chemical pollution, radioactive pollution, implementation of protective measures, the reasons for employment and lifting of the protective measures and the effects which may be expected both from the pollutant itself and the protective measures.

As mentioned above, the greater part of the information provided will reach the public through the media, so that allowance will have to be made for the possibility of distortion of even the most simple facts. Socio-economic and political factors might become the most important factors in the decision making process as was clearly experienced in the Soviet Union after the Chernobyl accident. If, however, the population believes that the protective measures are introduced with the prime purpose of protecting them solely against adverse health effects from the pollution, they may press the decision makers to reduce intervention levels even further. Statements must be prepared as part of the emergency plan to ensure that a clear and concise message can be sent immediately.

Communication must continue after the emergency so that long term consequences can be explained and rumours avoided. Credibility is vital and where possible communication channels should already be opened with an identifiable spokesman to maintain continuity. Reasons for lack of information should be given. In the field of Public Education action should be taken in close co-operation with professional groups, teachers, physicians, engineers and technologists since these can reach the public.

3.6. Protection and education of the Public

It is morally indefensible not to provide protection for the public against major hazards over which they have no control. However, the publics acceptance of protection measures is dependent on the level of understanding of the risks involved, the protection measures specified, the perception of risk and the balance of risk between adopting official advise and developing their own survival strategy. Trust is an important issue here. The credibility of a communicator and the message is critically dependent on the trust and belief in the organisation. It is also highly dependent on the different perceptions of risk. A message on the motorway to slow down to 30 mph because of fog may be completely ignored if the driver feels safe and in control. This is strongly reinforced if there is no immediate evidence of fog. Likewise an official message to remain inside and close and doors and windows in the event of a toxic gas release will not work if the receiver of the information perceives that it is safer to leave the area completely.

In a report [8] published by the Nuclear Energy Agency (NEA/OECD) radiation protection is described as a subject which is impenetrable to the layman and as wide as it is complex. For this and other reasons radiation hazards are perceived to exceed all others and the public appear to have a poor image of the radiation protection specialists. Radiation protection and nuclear engineering institutions provide support for education and training initiatives [9, 10]. In the European Union a Council Directive[11] on public information in the event of a radiological emergency was adopted in 1989. It defined action which must be taken when an emergency occurs but it also emphasised that the public should understand the issues involved and be prepared in advance of the emergency.

The Commission of the European Communities has undertaken a range of activities in support of public information including the publication of a video on ionising radiation [12] and a brochure [13] entitled "Radiation and You". The Commission also organised several seminars on Public Information and at a Seminar [14] held in 1988 it was suggested that teaching material on radiation protection should be developed for use in Schools of Member States. The Directorate General XI (DG XI) carried out the task [15] at Community Level. A first draft of the Teacher's Manual was developed and from this point the competent authorities of the member states contributed their invaluable support. Radioactivity, ionising radiation and non-ionising radiation are fairly complex and abstract subjects in particular for younger and less advanced pupils and so the option of a "spiral curriculum" was chosen. This means that items recur in a gradually more complex form. The material was set out in five levels covering ages 6 - 16. In the first three levels emphasis is placed on relating the pupils personal and every day experiences and they are made aware of the risks and benefits of ionising radiation. In the final two levels a more detailed examination is made of the subject from both a technical and social point of view. The original text was tested by the University of Utrecht in five member states. The reactions were generally favourable, a number of suggestions were made and the course thoroughly modified and edited into a convenient loose leaf format.

A test printing of the text was produced in 1993 and presented to a group of European Communities experts together with a formal presentation of the survey results. Reports were given by teachers involved in the testing and the two original authors responded to the comments presented their views. The course was modified in response to recommendations made at this meeting. New information coming from Chernobyl was added and potentially confusing material on the "green house effect" was deleted. The first edition in loose leaf format was published in 1994. This will not be the end the story; success will depend crucially on the sharing of the experience gained by teachers who choose to use this book.

3.7. Communicating the scale of an event

When an emergency occurs the authorities should issue a prompt report of the significance of the event in consistent terms. This again poses a communication difficulty, namely how to give a simple unambiguous message which conveys the urgency and seriousness of the situation. The nuclear emergency poses the most difficult challenge to communication and the International Nuclear Event Scale for Prompt Communication of Safety Significance [16] is now used for communications to the public about accidents or incidents at nuclear power plants. The scale was designed by an international group of experts convened jointly by the International Atomic Energy Agency (IAEA) and the Nuclear Energy Agency of the Organisation for Economic Co-operation and Development (NEA/OECD). The event scale classifies only the nuclear or radiological safety significance of the event on a scale of seven levels. The lower levels, 1 to 3, are termed incidents and the upper levels, 4 to 7, are termed accidents. At each level the scale is characterised by the off-site impact, the on-site impact and the degradation of defence in depth in the plant. The designers made the scale logarithmic so that the radiological consequences increase by a factor of 10 on each upward step as illustrated in Table 4. It should be possible to make an estimate in broad terms of the quantity of fission product activity released and thus assign a tentative level on the Scale. When the release is at levels 3 and 4, the relative radiological significance of the radionuclides is assessed by comparing the appropriate committed effective dose for intakes by all routes to the critical group.

The USA National Council on Radiation Protection and Measurements (NCRP) have issued a Commentary[17] to initiate a dialogue about what information and what form of presentation can prove helpful to the public in assessing an emergency situation with a potential for radiation exposure. NCRP recognise that a logarithmic scale can be used to show more detail at the bottom end where most actual exposures occur. They are aware that most people do not understand a logarithmic presentation. A further disadvantage arises because the scale is open ended (no zero) and length along the scale is not proportional to the effect. The same problem occurs in the Richter Scale index which uses a logarithmic scale for communicating the magnitude of earthquakes.

NCRP propose to use Index Values of radiation dose which they identify by giving Bench Mark values expressing the effective dose in appropriate units rounded to a simple integer value. For example, the radiation dose to the public at the Three Mile Island reactor accident was less than 0.01 mSv (1 millirem) and is assigned Index Value 1. The annual limit for radiation workers established by the NRC is 50 mSv [18], an Index Value of 5000 for one years exposure. The US Environmental Protection Agency (EPA) suggests a once in a life time lifesaving action level of 750 mSv for emergency workers [19] and Index Value of 75 000. It is recommended by the EPA that the emergency worker exposures be kept well below this level of exposure.

On this scale the average annual dose to the USA public from medical diagnostic procedures has Index Value 50, and the typical dose received by an astronaut on a "conventional" space mission has an Index Value of 500. The radiological consequence of INES Scale 3 would have an Index Value of between 10 and 100 and the consequences of Scale 4 would lie between Index Values 100 and 1000.

The graphic display of this scale presents practical problems since it must cover at least 5 decades. The NCRP propose to retain linearity and so they have used a "magnification" approach to show the detail at different levels. They recommend that research that evaluates the usefulness of the index should be given high priority. The NCRP also welcome comments on this approach from any reader of the Commentary.

4. ELEMENTS OF AN EMERGENCY INCIDENT

A recent study by Haines[24] on emergency management in the offshore industry interpreted the emergency response process as a dynamic risk management exercise. Fig. 3 shows the main components of the emergency management process for an offshore incident[25]. The four key components are identified as;

- Raising the Alarm
- Information gathering
- Team Strategy
- Mitigation and Control Actions

These are analogous to a risk management process (identify - assess - evaluate - reduce risk) and have the basic components of a human control system (sense - think - act).

***Alarm*:** Following an incident such as a fire or explosion, the installed safety systems will detect the presence of gas or fire and this will normally result in an audible alarm, backed up by public address announcements. This is the trigger for the incident control team to assemble and personnel to muster at the designated assembly points. On an offshore installation, there is no immediate escape beyond the limits of the installation, therefore there is a legal requirement to provide a temporary refuge (TR), usually the accommodation module, from which to control the incident and for personnel to muster. Demonstrating the integrity of the TR against smoke and gas ingress during a fire is now an important part of the Installations safety case.

***Information Gathering and Situation Assessment*:** The second stage in the process is to assess the situation. This is usually carried out in the incident control room (ICR) located within the temporary refuge. Often the ICR is the Process Control Room as this room provides the most rapid access to information on plant status and control systems. It is vital that enough information is gathered to support analysis and decision making. Without this the problem cannot be assessed and control will be impossible.

Typical Information required includes a preliminary assessments of the source of the incident, the damage inflicted, possible causes and consequences and the likelihood of escalation. In the early stages priority will be attached to accounting for personnel at muster areas. The greater the knowledge of the plant, its layout, functions, vulnerabilities, system responses, etc. the faster can the situation be appraised. This is an important pointer to education and training requirements. The ER Team must have detailed knowledge of the plant, processes and equipment etc. in order to function effectively.

***Team Strategy*:** Assessment is a continuing process which is revised and updated minute by minute and leads to the formulation of a strategy to control the incident and protect personnel. The primary objective is to protect life, with protection of assets and the environment as a secondary priorities. The team strategy will involve many considerations, for instance; deployment of fire fighting teams, shutting down the process and depressurising vessels, locating fatalities and rescuing injured personnel to the TR, deciding when and how to abandon the installation. There are other considerations to deal with apart from the immediate ones associated with the plant, for example, arranging off-site support, dealing with the media and the concerns of the family and relatives of personnel on Board.

The team strategy is derived from and dependent on the information gathered and prevailing situation. In some cases the strategy is laid down as a standard emergency procedure but there are enormous uncertainties associated with major accidents and the response chosen will ultimately depend on the particular situation and conditions at the time. This is where professional knowledge and clear thinking are required. When the normal systems have been disturbed the ER Team may well need to revert to deeper levels of understanding of the processes (knowledge based reasoning) and not rely on standard operating operating procedures and rules.

Clearly, there is ample room for human error to be made with significant implications. For instance a decision not to abandon the platform, at a time when abandonment was possible, followed by a worsening of events which prevents platform abandonment could lead to massive loss of life. Therefore, early evacuation of all non essential personnel (those not directly involved in controlling or managing the incident) is normal procedure.

***Mitigation and Control*:** The actions taken are totally dependent on the strategy which has been formulated. Many of the control actions are automated e.g. actuation of emergency shutdown valves, activation of fire and gas protection systems, and do not require human intervention. However, there are many manual actions which may need to be taken to reduce the risks. Some manual control and mitigating actions may be clear and unambiguous Unfortunately this is not always the case. Some actions will be open both to errors in interpreting the strategy and errors in actually implementing the required action.

Command and control during a major accident are vitally important during the assessment phase and particularly during the mitigation and control phase. In difficult and dangerous conditions, the response team as well as non essential personnel need strong leadership from the Emergency Director . In the Piper Alpha disaster the OIM did not provide adequate leadership and although in that particular case this may not have contributed significantly to the end result, there is clear evidence from the armed forces and the emergency services that strong leadership is essential. This has now led the offshore industry to require competence assessments for all Offshore Installations Managers.

5. TRAINING OF EMERGENCY TEAMS

The key objective of an emergency management system are *to prevent or reduce the likelihood of consequential loss in the event of an emergency occurring.*. The perceived benefit of training is improved performance in an emergency, i.e. an increase in the likelihood of successfully controlling an incident and minimising losses. The cost of training of emergency personnel, however, is considerable [20] and may consume as much as 5% of manpower resources on a nuclear site. Teams should be trained together and practice together. Exercises can be used to maintain skills and can also be used to test the adequacy of plans. Every drill or exercise must be observed and critically discussed. Unplanned tests are not recommended. The aims of the exercise should specify the priorities in training and testing. All participants have a duty to remain prepared for any emergency using the following guidelines:-

(a) Know what is contained in the emergency plan,

(b) Know your own role,

(c) Know the site procedures,

(d) Train in your own task,

(e) Familiarise yourself with the equipment,

(f) Get involved in rehearsals and drills,

(g) Be aware of help available outside your own organisation.

5.1. Drills

Drills are used for exercising procedures and are specially useful in the early stages of an emergency e.g. mustering of personnel and use of protective equipment following an alarm. Individual skills in the use of emergency equipment e.g. breathing apparatus must be maintained by drills. Team drills are most important as there are few activities associated with emergency response that do not benefit from drills. Every effort should be made to simulate the pressure on time and resources which would occur in a real event. Drills and training improve the speed of response to an alarms and with sufficient training the response may become efficient and automatic. However, there will always be some accidents where the standard procedures do not apply.

5.2. Exercises

Exercises are designed to test an emergency plan or provide training for emergency response teams. These are mandatory now in a number of industries. Exercises can be carried out with varying degrees of realism and a number of commercial organisations provide emergency management training courses either on-site or at the training centre facilities to meet this need. It is arguable that the more realistic the exercise the better is the test and the training quality. In some cases an attempt is made to test the full scope of the plan on site. In the nuclear industry it is difficult to justify the use of real reactors and live radioactive sources and so some exercise artificiality is inevitable. Similar constraints apply to any other exercise with toxic substances. However, the provision of a realistic simulated environment such as a control room with visual and audio systems to mimic the real plant, can provide sufficient realism for the trainees to respond as if the exercise were real. In the offshore industry the competence of the OIM can be assessed using such simulators. These test the candidates competence to manage "under fire" for a number of major accident scenarios. For on site exercises, the appointment of an Exercise Director and a Safety Officer is essential to success and safety. A communication boundary must be established to prevent the escape of exercise messages.

In an emergency exercise, individuals should perform their normal task but some role playing will be necessary to enliven the scenario. Participation in the exercise may be restricted to groups or teams with a specific task such as media contact and rumour control. It is rare for exercises to involve real members of the public and media. The participation of operational staff who have primary responsibility for safety of the plant presents some difficulty since these staff are dedicated to preventing the event which they are invited to simulate. It is recommended that the their participation should be conducted on a simulator so that the safety of the plant is not compromised.

It is the experience of emergency management trainers that the performances of individuals and teams in simulated emergencies, improve with practice and experience e.g. during courses and over a number of exercises. This suggests that cool heads in an emergency can be acquired. However, the degree of correlation between individual/team performance under simulated circumstances those achieved in a real emergency is difficult to assess objectively and more research is needed in this area.

***Table Top Exercises*:**

Exercises conducted at a table, "table top exercises", [21] are the most economical. They may be classified [22] as follows:-

(a) **Linear Exercises**. The scenario does not deviate from a predetermined sequence of events and the participants follow a route for which correct solutions have been specified. Inexperienced staff or students can perform directing staff duties. The method can be transferred internationally with few language difficulties.

(b) **Open Exercises**. The exercise is free running from the point of the initial scenario and the out-come is not known. It is enjoyed by students but demands considerable directing staff time. It is particularly useful for developing communication procedures between on-site and off-site facilities. A bank of messages, media questions and the injection of additional scenario items is required to keep up the momentum.

(c) **Communication Exercises**. In many exercises communications is the only part of the emergency scheme to be tested and the table top exercise is ideal for this purpose. In some cases a field exercise may be enhanced by using a table top team to provide communication responders. This avoids the use of dedicated networks and reduces the chance that exercise messages may be received outside the area of play.

(d) **Committee Exercises**. When broad issues are on the agenda and a large numbers of players must participate the exercise can be conducted like a committee meeting at which regulatory authorities, service organisations etc., are represented. A committee exercise encourages an exchange of views and a good chairman can ensure useful conclusions. The NRC conducted such an exercise with 90 players (23) which ran for two days.

5.3. Workshops

The exercises described above are run under pressure. This can be reduced by using a workshop format so that the participants may work in groups. The groups may synthesis a plan from elements provided by the directing staff. Alternatively the plan may be analysed to find out if it is sufficiently robust to cope with problems such as an overload of the communication systems or with an extended emergency.

6. STRESS AND HUMAN FACTORS

When an incident initiating event occurs, the risks to the work force increase immediately by several orders of magnitude. The knowledge of this risk will increase the level of stress on the individuals exposed to the danger. Psychologists have long known that stress is an important performance shaping factor and can both enhance and reduce the performance of humans. Fig. 4 (28) shows in qualitative terms the expected response. As the level of stress increases, the performance of individuals at first increases as a positive response to challenge. However, as the stress increases further the performance peaks and further increases cause a reduction in performance until, at extreme levels of stress, the performance almost falls to zero as feelings of hopelessness and inability to cope invade the mind. Psychologists believe that the probability of error under exceptionally stressful conditions can lie between 0.5 and 1.

It is clearly important that the emergency response team and particularly the overall manager of the emergency must achieve a very high level of performance during an incident to stand the best chance of survival. This can be achieved in two ways. Firstly, because it is recognised that stress-performance profiles varies enormously between individuals, then it makes sense to select individuals which have a high stress tolerance level. This is the basis of the Offshore industry requirement for OIM competency tests. The alternative is to reduce the level of stress perceived by the ER team and this is discussed below.

To reduce stress it is first necessary to understand the causes of stress. A list of factors which promote the occurrence of errors are listed in table 4 (29). When the consequences of error are high, these factors are the major contributors to stress. The error promoting conditions which have the most significance in an emergency are:-

CATEGORY	RISK FACTOR	MANAGEMENT IMPLICATION
Unfamiliarity	17	Education Training and Procedures
Time Shortage	11	Incompatible goals, Education and training
Noisy signals	10	Communications
Poor system /Human user interface	8	Design
Information Overload	6	Design, Communications
Poor Feedback	4	Design, Communications
Inexperience	3	Training
Poor Instructions or procedures	3	Training, Procedure

In effect, the probability of error will be highest when deal with unfamiliar situations, in which there may be a high level of disorganisation and complexity, in which there is reduced ability to control events with limited access to information, diminishing options for a solution and limited time to deal with the problem.

7. TRAINING AND RESEARCH IMPLICATIONS

The various models and descriptions covered in the paper provide a basis for selecting topics for inclusion in education and training courses. They also provide indicators for where future research would be useful.

Clearly there are a wide range of topics which could and should be included in a syllabus for emergency management and the content and format would have to be tailored to particular user. However, some useful topics for inclution in a syllabus for process workers is listed in table 6 and an extended list of topics for more detailed educational use is listed in table 7.

Specialised training and facilities are required to meet the needs of emergency response personnel in those industries where majoraccident hazards are present. Incident Managers with high stress tolerance levels should be selected as they are more likely to perform better under highly stressful conditions. Specially developed simulators already exist and Virtual reality software is also becoming available. These types of facility would help with the competence assessments of Incident Managers and On-scene commanders by increasing the perception of real risks and dangers, without the effects of physical danger. However, the possibility of psychological trauma remains. Such facilities provide useful experimental material for industrial psychologists.

The perception of risk during an incident is more important than the actual risk. The belief of the individual in the emergency management system's (facilities and organisation) ability to control, to protect, to provide options and ultimately provide a means of evacuation or escape will reduce stress and promote high performance. This depends on the culture of the organisation as well as on the individual characteristics of personnel. However, education and training will contribute significantly to stress reduction by improving every individuals perception of their ability to cope.

Risk Management action prior to an event can significantly reduce stress in the ER Team and other personnel on an installation by providing a system which has the effect of reducing:-

(a) The personal consequences of human error during an incident,

(b) The likelihood of error.

These two factors are in fact linked since a reduction in threat to life, will decrease the stress and the decrease the of probability of errors.

Personal consequences can be reduced by providing options for protection (e.g. temporary safe refuges) or for evacuation and escape in the event that evacuation becomes a necessity. This can only be dealt with prior to an accident and forms the basis of much of the Safety case work which is now required by industry. However such analyses form an important part of emergency planning.

Unfamiliarity has the most significant effect on error probability. The likelihood of errors can be reduced therefore by training personnel to become familiar with the rare events. Through exercises and by training staff to be familiar with emergency procedure, plant, processes and layout etc. The use of exercises and simulators with high levels of realism improves the training. However, to be effective the exercises need to be repeated at regular intervals. Research to quantify the effectiveness of training as a preparation for a real emergency would be useful.

Every effort should be made to increase the time available for dealing with an emergency. Time stress increases with the mismatch between the time available for emergency control action and the time required for action. The time available will be dominated by the particular characteristics of the incident e.g. rate of escalation. However, the time required for problem solving can be improved to some extent by education and better understanding, by training and practice, as well as by developments in efficiency enhancement tools e.g. special systems to record events, graphical display tools to improve the generation of the big picture, Computerised risk prediction systems e.g. to predict dispersion and spread of toxic or radioactive plumes.

8. REFERENCES

1. Nuclear Regulatory Commission, USA.; Planning Basis for the Development of State and Local Government Radiological Emergency Response Plans in Support of Light Water Nuclear Power Plants, NUREG - 0369, EPA 520/1-78-016 Reprinted 1988.
2. Nuclear Regulatory Commission, USA.; Criteria for the Preparation and Evaluation of Radiological Emergency Plans and Preparedness in Support of Nuclear Power Plants. NUREG - 0654 / FEMA - REP - 1, 1980.
3. Emergency Preparedness - Planning for the Unexpected. Courses provided by Institute of Nuclear Power Operations, 700 Galeria Parkway, Atlanta, Georgia, USA.
4. An Engineer's Responsibility for Safety, Proposed Syllabus for an Undergraduate Awareness Course, Hazards Forum Document, 1992, Institute of Civil Engineers, 1 Great George Street, SW1P 3AA

5. G D Kenney; The Management of Safety and Emergency Planning, Hazards Forum Conference 12-13 October 1993, The Successful Management of Safety, Institution of Mechanical Engineers, London, 1993

6. Thomas E Drabek; Strategies Used by Emergency Managers to Maintain Organisational Integrity Environmental Auditor, Vol 1, No.3, p. 139-152, Springer-Verlag 1989

7. Thomas E Drabek; Emergency Management - Strategies for Maintaining Organisational Integrity, Springer-Verlag New York Inc. 1990.

8. Nuclear Energy Agency/OECD; Nuclear Energy : Communicating with the Public, ISBN 92-64-13456-5, OECD Paris, 1991.

9. LAKEY, J.R.A.; The Rôle of IRPA in the Advancement of Radiation Protection, Proceedings of the IV National Congress of the Spanish Radiation Protection Society, Salamanca, 26 - 29 November 1991 p.59-65, CIEMAT, Madrid, 1992

10. LAKEY, J.R.A. editor with others; Off-site Emergency Response to Nuclear Accidents, Manual based on training courses organised in 1991 and 1992 at the SCK/CEN Mol, Belgium. Publications of the Commission of the European Communities, Radiological Protection No 60, Luxembourg 1993.

11. Commission communication on the implementation of Council Directive 89/618/Euratom of 17 November 1989 on informing the general public about health protection measures to be applied and steps to be taken in the event of a radiological emergency; 91/C103/03, Official Journal of the European Communities C-103.

12. Radiation Types and Effects, Origins and Control Sheffield University Television, Catalogue Number CC-ZV-92-902-EN-V

Radiation and You; Commission of the European Communities, ISBN 92 825 9486 6, Catalogue Number CC-54-88-053-EN-C

13. Radiation Protection and Training for Workers; Radiation Protection No 45, EUR 12177, Commission of the European Communities, Luxembourg 1989

Radiation and Radiation Protection a course for primary and secondary schools, Radiation Protection No. 67, ISBN 92-826-6731-6, Commission of the European Communities, Luxembourg 1994.

see also

LAKEY, J. R. A., DRAIJER, J. G. H. and TEUNEN, D., The European Commissions's Teacher's Manual on 'Radiation and Radiation Protection' - a Course for Primary and Secondary Schools. 19th Annual Symposium, The Uranium Institute, 7 - 9 September, London 1994

14. IAEA and NEA, The International Nuclear Event Scale, Users Manual, Revised and Extended Edition 1992. IAEA/INES/92/01, IAEA, Vienna, September 1992

15. National Council of Radiation Protection and Measurements. Advising the Public about Radiation Emergencies (1994), NCRP Commentary No. 10, ISBN 0-929600-38-X , NCRP Bethesda, Maryland, USA

16. US Nuclear Regulatory Commission (1990), Standards for Protection against Radiation, Nuclear Regulatory Commission, 7590-0-1, 10 CFR 20 Final Rule, effective date 1 Jan, 1992, US Government Printing Office, Washington, DC, USA.

17. US Environmental Protection Agency (1990). Manual of Protective Actions for Nuclear Incidents, Report EPA 0520/1-75-001-A, US Environmental Protection Agency, Washington, DC,USA

18. SEDGE, L.A. and LAKEY, J.R.A., Fire Training, Drills and Exercises in the Nuclear Industry, The Nuclear Engineer, vol. 28, 16 - 20, 1986.

19. LAKEY, J.R.A., BARRATT, K.L. and MARCHANT, C.P, Nuclear Reactor Emergency Exercises and Drills, IAEA-SM-280, Rome, 1983.

20. LAKEY, J.R.A., A Table Top Exercise and Workshop, Eighth International Congress of the International Radiation Protection Association, Montreal, Canada 17-22 May 1992.

21. ZECH, G., 1990.; Post Emergency Table Top Exercise, Lessons Learned Report, NUREG - 1441 / FEMA - REP - 16.

22. FOSTER H. D. Disaster Planning, New York, Springer-Verlag 1979.

23. HAINES J. Emergency Management and Incident Control on Offshore Installations, MSc Thesis, Cranfield University, 1995

24. HAINES J and STRUTT J E, "Emergency Management and Incident Control on Offshore Installations" to be published.

25. OTWAY H J and WYNNE B ; Risk Communication: Paradigm and paradox, *Risk Analysis* 9, 141, 1989

26. KASPERSON R. E. , KASPERSON J X and RENN O; The Social Amplification of risk: Progress in developing an integrative framework. In *Theories of risk* Ed. S Krimsky and D Golding, New York 1992

27. SWAIN A. D. and GUTTMAN H. E. Handbook of Human Reliability Analysis. NUREG/CR - 1278 1980, Sandia Laboratories Albuquerque New Mexico.

Table 1 Recent major disasters

SECTOR	ACCIDENT	Fatalities	DATE
Nuclear Accident	Chernobyl, USSR	31	25/04/1986
	Three Mile Island, USA	Nil	28/03/1979
Offshore Accident	Piper Alpha, North Sea (UK Sector)	167	06/07/1988
	Alexandre Keilland, North Sea	123	27/03/1980
	Sea Crest, Thailand	91	11/03/1989
	Glomar Java Sea, China	81	26/08/1983
	Pohai 2, China	72	25/11/1979
Chemical Industry	Bhopal, India	> 2500	03/12/1984
	Flixborough, UK	28	/06/1974
	Seveso, Italy	Nil	10/07/1976
	Mexico City, Mexico	> 500	19/11/1984
	Basle, Switzerland	Nil	25/04/1986
Air Accident	Challenger space shuttle, USA	7	28/01/1986
	Tenerife, Spain	583	27/03/1977
	Nagoya Airport, Japan	264	26/04/1994
	Manchester Ring Way, UK	55	22/08/1985
	Amsterdam, Netherlands	204	04/10/1992
Shipping Accident	Estonia "ro-ro" ferry	> 900	28/09/1994
	Herald of Free Enterprise	188	06/03/1987
Railway Accident	King's Cross Underground fire, UK	31	18/11/1987
	Clapham Junction, UK		12/12/1988
	Mississauga, Canada	Nil	11/11/1979

Table 2 Important elements of an emergency plan

- The type of accident which might occur
- The organisations involved in the emergency management and the key personnel
- Communication links telephones, radios etc.
- Fire fighting equipment, damage control and repair
- Technical information e.g. chemical and physical characteristics and dangers relating to the substances involved.
- Physical Information relating to the layout of the plant, pipelines and safety critical components such as emergency shutdown and venting equipment etc.
- The emergency procedures and evacuation arrangements to be used
- Contacts for obtaining advice on e.g. meteorological conditions, emergency transport systems, first aid , hospital services
- Arrangements for dealing with the press

Table 3 Management of nuclear emergencies: Important organisational elements

Command and Control	Overall command and local command must exist and be accountable. Control and communications must be disciplined. The Emergency Director should have the task and the authority to co-ordinate, supervise and direct the action of all organisations involved locally in the protection of the population.
Alarm, Alerting and Notification scheme	Early notification is important and mandatory in most cases. The correct classification, correct priority, correct point of contact are equally important. Early messages should be in a standard format and communication procedures should be drilled. Response to alarm signals should be automatic.
Assessment	Initial assessment is derived from plant parameters. The on-site and off-site radiological data may be interpreted with "real-time" computer codes. These activities rely heavily on engineering judgement.
Public Information	In the European Union this is mandatory. Communications to local authorities is most important. Information is also released through press conferences and links with local organisations and voluntary groups. Rumour control relies on the monitoring of incoming calls from the public and co-ordination with press calls of other participants.
Radiological and chemical Protection	This includes assessment and control, authorisation of excess exposures, records, contamination control and decontamination
Public Health and Safety	It is essential to give information to public authorities about the potential hazard and the proposed protective actions. Use of media must be co-ordinated from a single source.
Social Services	The welfare of groups with special needs, the elderly and all kinds of hospital and sheltered institution.
Environmental Survey	Telemetry or mobile monitoring must be calibrated against traceable standards. Informal data should be used only if calibration is available.
Fire and Rescue	On-site and off-site fire service should use common training standards, communications and equipment.
Coastguard	Assistance and control of access by water.
Law Enforcement	Site security and public order. Police are in Command and also provide control through communications together with transport and access control.
Transportation Services	Both private and public services are required for movements of emergency workers and the public.
Recovery Teams	Following engineering assessment off-site resources are required for recovery and would be expected to work under normal health physics control

Table 4 The radiological aspects of the INES scale

INES Scale	Release Iodine 131, (tera becqueral)	Dose to the Public (millisievert)
7	> 10,000	
6	1,000 - 10,000	
5	100 - 1,000	
4	< 100	1 - 10
3		0.1 -1

Table 5 Error Promoting Conditions

CATEGORY	RISK FACTOR	MANAGEMENT IMPLICATION
Unfamiliarity	17	• Education and Training • Procedures
Time Shortage	11	• Incompatible goals
Noisy signals	10	• Communications
Poor system /Human user interface	8	• Design
Designer/User Mismatch	8	• Design
Irreversibility	8	• Design
Information Overload	6	• Design • Communications
Technique Unlearning	6	• Education and Training
Knowledge Transfer	5	• Education and training
Misperception of risk	4	• Training
Poor Feedback	4	• Design • Communications
Inexperience	3	• Training
Poor Instructions of procedures	3	• Training • Procedure
Inadequate checking	3	• Organisation • Maintenance • Procedures
Educational Mismatch	2	• Training • Organisation

Table 6 Topics to be included in emergency management training for process plant workers

• Safety awareness
• Process technology
• Accident prevention
• Compliance with regulations and company policies
• Corporate and supervisors liabilities
• Impacts of new legislation and regulations
• Working with regulatory bodies
• Emergency response plans
• Community support needs

Table 7 Topics for inclusion in education and training programmes on emergency management

Industrial Hazards	chemical and physical characteristics and dangers
Technical Information	Plant drawings, Layout of plant, piping and utilities, P&I diagrams, process diagrams etc.
Risk Analysis	Identification and analysis of major accident hazards, Prioritisation of emergency resources
Protection	Protection measures on and off site: temporary refuges, radiological and chemical Protection etc.
Emergency equipment	fire fighting, smoke hoods etc.
Emergency systems	ESD, Fire & Gas systems, venting systems, water deluge systems, emergency lighting and power etc.
Emergency plans and procedures	
Alarms and notification schemes	
Emergency Management	Command and Control, organisational structures, key personnel during incidents
Analysis and Assessment during incidents	Tools, procedures and technical aids to generate the Big Picture.
Additional information	Meteorological information
	Risk communication, social amplification risk
Evacuation Escape and Rescue	Analysis methods, in practice
Dealing with Outside Agencies	fire service, first aid, ambulance and hospital services, social services, press, relatives etc.
Accident investigations	and post incident analysis
Communications:	Communications theories informing the public, communications systems, messages and practical use
Human Factors	Causes and Effects of Stress, performance shaping factors

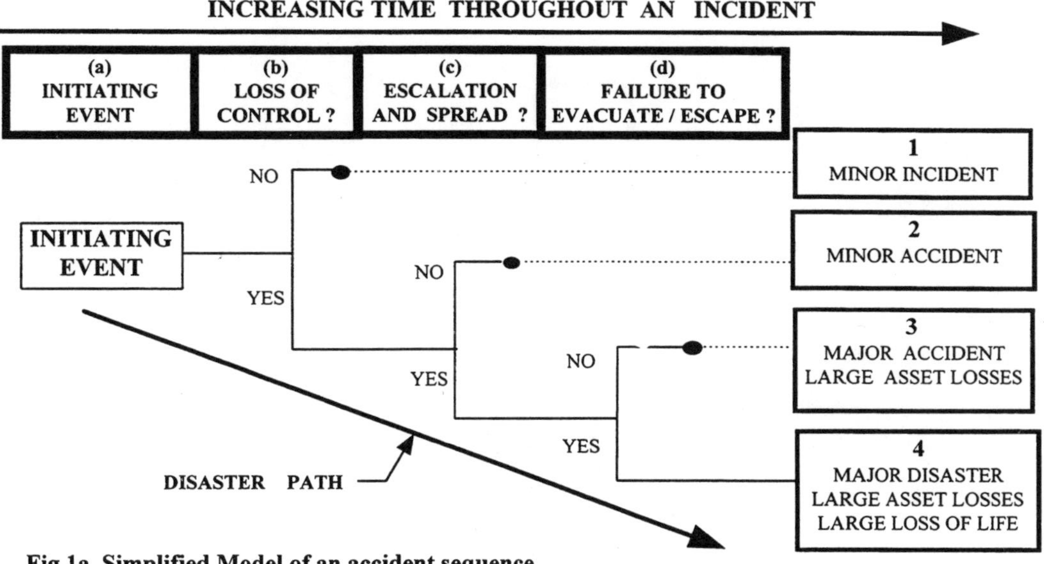

Fig.1a Simplified Model of an accident sequence

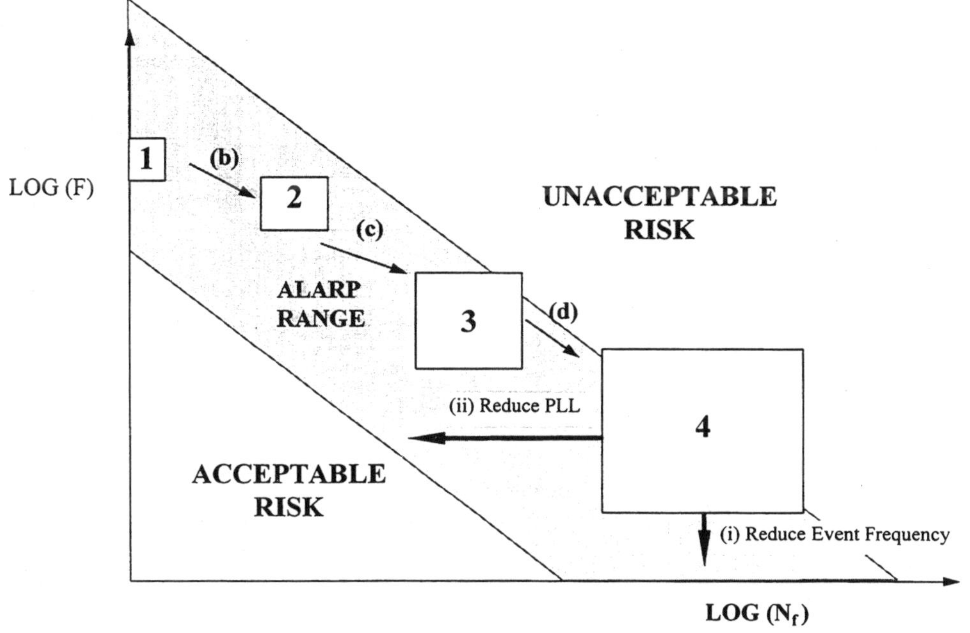

Fig.1b Risk Diagram corresponding to Fig. 1a

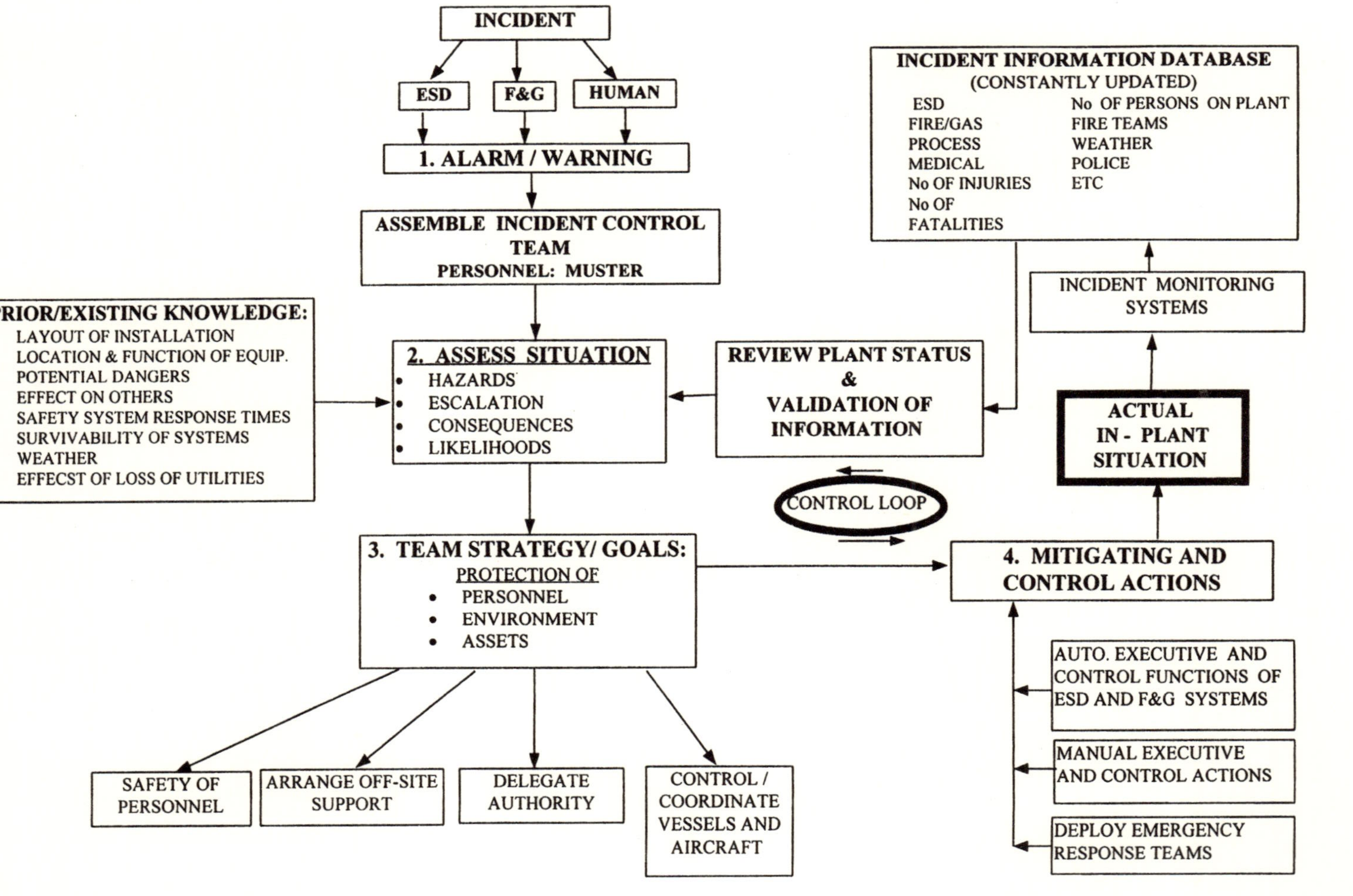

Fig. 2 The four key stages in Offshore incident management
(1) Alarm (2) Information Gathering (3) Team Strategy, (4) Mitigation and Control Actions

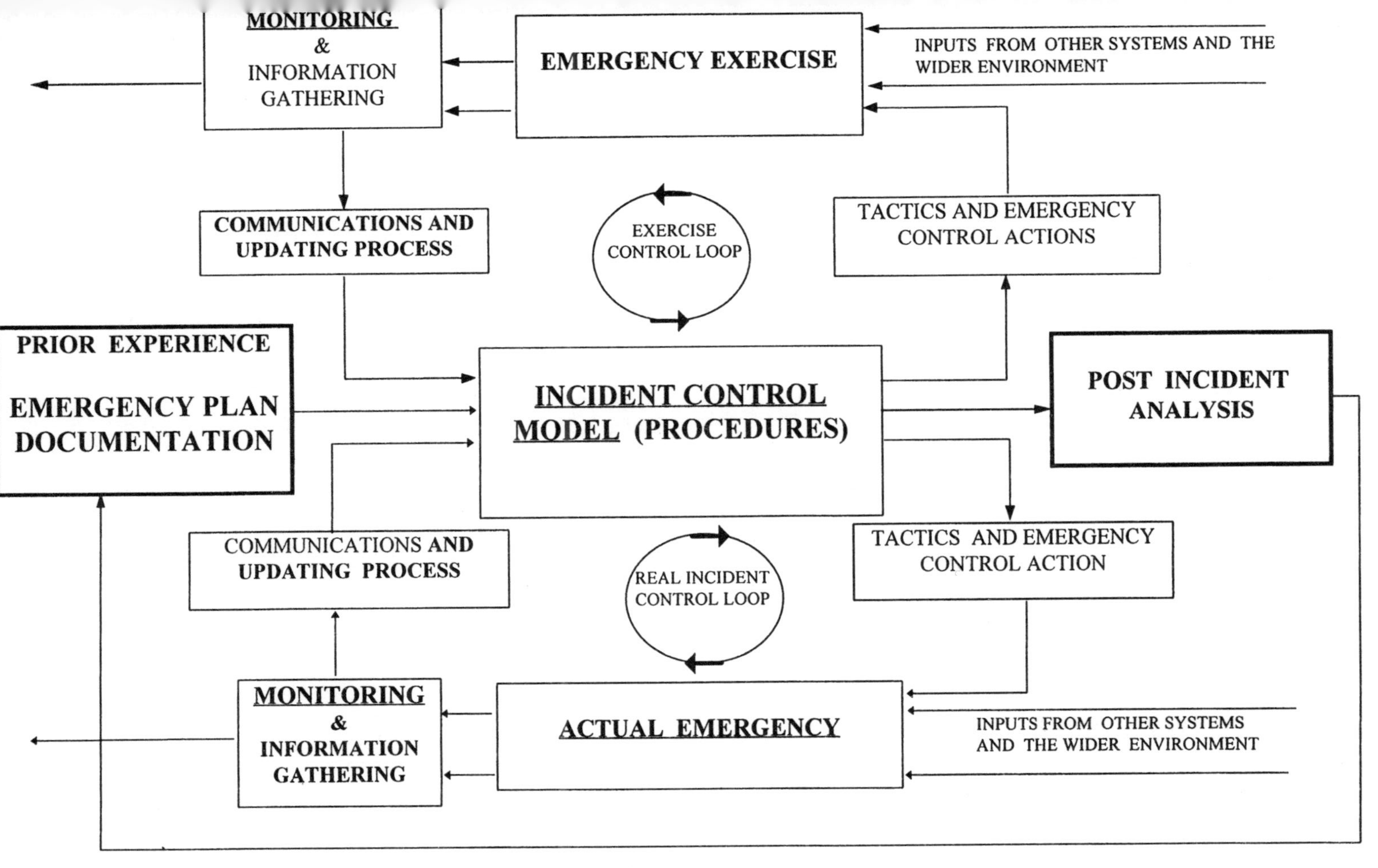

Fig. 3 Relationship between the emergency plan, actual and perceived situation

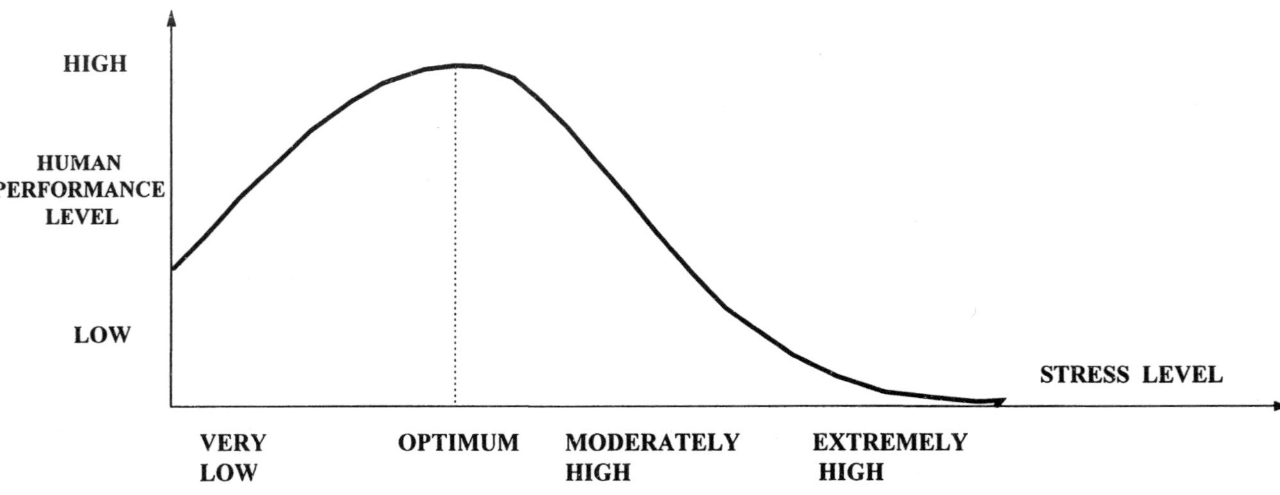

Fig. 4 Schematic representation of variation of Human Performance with level of stress

C507/010/95

Practical emergency management

E S J LARKEN Rear Admiral, DSO
Operational Command Training Organisation Limited, UK

SYNOPSIS

The paper examines the realities imposed by crisis and emergency management. Its focus is upon major hazard plant. It is rooted in current practical experience, and draws in particular upon the offshore oil and gas industry. It will identify, with practical examples, key factors in successful emergency management. These will include their application to the selection and development of persons in charge and their teams. Relevance extends to emergency prevention.

KEY WORDS

Crisis Management; Emergency Management; Emergency Prevention; Emergency Command; Command Teams; Command Training; Leadership; Leadership Training; Control of Risk in Major Hazard Plant; Command Appraisal and Assessment; Offshore Installation Managers (OIMs); Platform Managers; Major Hazard Plant Managers.

INTRODUCTION

"Crisis" - derived from ancient Greek word for *"Decision"*

The Oxford Dictionary defines an emergency as *"A sudden juncture needing prompt action"*.

"We know from experience that the circumstances of any major accident never repeat themselves". Brian Appleton, ICI and Assessor on the Piper Alpha Public Enquiry.

Despite efforts to the contrary, emergencies continue to outflank attempts at prediction, anticipation and prevention. Accidents, emergencies, and crises present by their nature an apparently endless ability to surprise. Examples include the Argentine invasion of the Falklands, the collapse of the Soviet Union, the re-election of the Conservative Party at the last General Election, the recent Kobe earthquake, the Piper Alpha disaster, the Kings Cross and the Bradford stadium fires and the loss of *Herald of Free Enterprise* and *Estonia* - the latter two, you will recall, being repetitions of type. In the current offshore context, reported gas releases number around 100 a year, of which some 10% are of a size equivalent to the initiating event of Piper Alpha.

This talk deals with the topic of emergency management, command and control. It will draw upon four years of training managers of major hazard facilities in the onshore and offshore oil and gas industries and the nuclear industry. Additionally it is underpinned by 33 years of service in the Royal Navy, of which 15 years carried direct involvement in surface and submarine command - eg. a key ship and task-group command role in the Falklands campaign and four years developing command training including supervision of all submarine advanced training and the 'Perisher' course.

I offer my view here as a practitioner, not a psychologist nor a human factors specialist, although I will draw from time to time upon some of the leading academic work in the field of emergency command.

The major hazard manager's role in emergency involves "command".

None have put this better than Lord Cullen in his report on the Piper Alpha disaster. Referring to the plant manager (known as the Offshore Installation Manager, or OIM), Lord Cullen said (Cullen Report para 20.59) that this post "calls for decisions which may make the difference between the life and death of personnel" and that "it demands a level of command ability which is not a feature of normal management posts". These responsibilities are increasingly vested in law upon line management for major hazard plant. So far as the offshore oil and gas industry is concerned, the relevant legislation is to be found in the new Prevention of Fire and Explosions, and Emergency Response (PFEER) Regulations, which have reached the Statute Books. Explicit, as in the nuclear licensing regime, is the need for sound emergency management.

MANAGING AN EMERGENCY

Offshore Installation Manager (OIM) in action provides an example.

As a mechanism for describing what I want to share with you about emergency response in major hazard plant, I am going to invite you to consider a structured example of an offshore emergency. We have a large, fixed platform typical of those found in the northern North Sea. It comprises (Fig 1 - at end of paper):

- an accommodation module, set above some services plant and topped by the heli-deck;
- a drilling and well-head module, with drilling derrick; and
- a series of oil and gas processing modules of compact layout, designed to resist fire rather than explosion, with associated gas and oil pipelines to and from the sea bed.

Our emergency scenario arises in the middle of a dark and, fortunately, calm night. It starts with a small gas escape in the production module which, shortly after detection, ignites and gives a modest explosion. The explosion ruptures the water deluge pipework and the crude oil pipeline between the 2nd stage separator and main oil line pumps. It causes some other collateral damage to fire walls. A large pool fire begins. Three men were working in the module.

The first need is for the whole organisation to respond: change gear, find out what is going on whilst taking initial actions, and to look ahead; the OIM must strive to stand above the detail.

The OIM is summoned from his bed by the gas alarm to the emergency command centre, there to assume command of the emergency support team. He arrives, very much wide awake, within two minutes of the alarm. His task is to change an unexpected, adverse, unstable and fast-moving situation into one that is at least contained, stable and controlled. The OIM must create and sustain coherence out of threatening chaos. This requires actions. Actions demand decisions. Let us take the OIM's perspective.

There is instantly a flurry of information and a mass of issues to consider and arrange into a mental picture of the emergency (for a schematic of developments and likely immediate issues (see Fig 2 at end of paper). From this mental picture, the OIM needs to take decisions to initiate actions. A side-effect of these will be more information (probably of improving quality), and requirements for ensuing decisions.

It can be seen already that during these first few minutes the picture is growing complicated. There is a great deal of communication, and from it much information to be processed. This of itself brings pressure. Add to this the pressure of uncertainty and concern for the outcome. Time constraints will aggravate sensations of stress impinging upon one and all. What about stress? Stress is for many a stimulus, at least for a while; some however it will hobble from the start. For everyone it will disturb (and differently for each) the natural sensation of the passage of time.

In this situation, the OIM must maintain control of him/herself and the team sufficiently to fulfil effectively the command role. Take a few of the OIM's essential competence credentials:

- to ***stand above*** the detailed tactical decisions (eg: fire-fighting; the personnel muster) and ***develop*** a strategic over-view;
- to ***make an early appraisal*** (outputs to include: worst credible course of incident; most likely courses; major measures to be set in hand and external resources activated);
- to ***create initial direction***, by making objectives clear; and to have ***set in hand*** the provision of on-platform and external resources;
- to ***monitor*** key details to ensure consistency with strategy (eg: plant shut-down and de-pressurisation; security of people; fire teams (effect v risk); effective information management and presentation; logistic management; practicable evacuation);
- to ***project ahead*** in the most systematic manner practicable, ***refine and continue to articulate objectives*** (life first, then plant) (to be repeatedly re-visited) and ***make good decisions***.
- to ***draw advice*** from the emergency command team and wherever appropriate;

The nature of this major hazard plant emergency is that all these facets are relevant. They are subjects of interest in their own right, but for this paper I shall concentrate upon command.

Certain core issues of command prove to be the most elusive, but I will offer you the analysis of a practitioner.

During the early stages, the OIM's actions will probably lie within the envelope of prescribed procedures. Operators respond in the main on the basis of procedural reflex. Much of the shutdown is automated. Personnel will follow muster procedures. Fire teams will assemble and make preliminary assessments. The process team will verify the shutdown and blowdown functions.

The emergency begins to move outside pre-planned responses and the operational envelope.

Very soon however, this emergency (as so many) begins to move outside the operational envelope. Any who doubt this expectation should consult the well-analysed records of civil aviation incidents. There may be errors. Muster areas may become smoke logged. The helideck may also be unusable on account of smoke. Weather may limit lifeboat availability, or in extreme cases make lifeboats virtually unusable. I myself have spent two hours in the middle of the night in a life-boat on the upwind side of platform, supposedly ready to go in a Force 9 under the command of an unbriefed and terrified coxswain. In one recent instance of gas release, within the first moments the Control Room Operator had inadvertently overridden the emergency shut-down system and isolated the fire water pumps - whereafter the gas cloud quickly entered the gas turbine intakes, but without an explosion. Yes, when things happen suddenly and stress impinges, even experienced people can act in illogical ways. When, as so often, we get outside the operational envelope, the OIM needs more than ever to diagnose correctly, project ahead, act decisively and improvise. As I have said elsewhere and in another context, he will in large degree then be operating "in unknown territory without maps". This demands the utmost resources of courage, coolness, skill, and background knowledge.

We should therefore examine the core command activities which take place as we move outside the procedural envelope. These are: correct diagnosis; clear objectives and creation of direction; future projection; time management; team management; and overall control; all with monitoring feed-back loops. Let us consider some elements of these.

Objectives and direction are essential but elusive.

Following a correct, perhaps multi-faceted, diagnosis, creating direction is an elusive concept; yet it is essential to understanding the role of the emergency command function. It involves the skills of ***strategy formation*** and ***decision making***, both of which must take place under ***conditions of time pressure and limited information*** in a situation which may be moving ever further ***outside the operational envelope***. It involves, first and foremost, the establishment of objectives, which must be clearly articulated to the team.

> In this case, the OIM after some minute or so feels sufficiently clear about the situation that he calls firmly for attention (with the secondary purpose of ensuring everyone has noted his arrival) and makes a brief statement of the situation as he sees it. No-one disagrees. He states his objectives, turns to the Public Address (PA) system and makes a firm and factual, yet warm, broadcast to all platform personnel.

To develop the concept further, we must deal with the factors highlighted above, starting with decision-making.

Decision making - there is no quick fix nor easy answers

Decisions operate, self-evidently, at both strategic and tactical levels. Derived from the experience of practitioners, there are two principal genres of decision-making model available from psychological research. Both, if they are to make sense, need to carry clear objectives, themselves subject of decisions. The first approach can be termed naturalistic (intuitive); the second classical or analytical (thorough and long-winded). For the detail, please see references (JSP 101, Klein and Orasanau).

Naturalistic decision-making theories have been developed from observation of, for example, expert fire team commanders with long experience of practical command skills. It involves recognition of the nature of the situation and determining actions from intuitive mental modelling - with minor modifications - to achieve a sufficient and satisfactory outcome. The key point here is that this work has been done observing practitioners of long-proven ability in an area which, whilst demanding in terms of functioning under pressure, lies within their operational envelope.

This approach may give you difficulty in the industry context. Here we shut down plant and make safe when real trouble threatens. Plant managers do not spend their lives preparing, primarily, to fight serious emergencies. They do so at a major level (e.g., outside the operational envelope) possibly once or twice in their careers - perhaps never. For the industry emergency manager therefore, naturalistic decision making, where the first apparently workable option is adopted, offers much scope for false judgements, and for rapid disappearance down metaphorical rabbit-holes.

Therefore, in the complex emergency situation on a major hazard site, the emergency commander needs to strike a delicate balance between the temptation to adopt intuitively the 'obvious' solution and an analytical approach which questions and tests whether or not the solution is right - and to be prepared to change direction if necessary.

Relate this to our incident. Naturalistic decision-making has carried us safely through niceties of shut-down and blow-down; the conduct of the muster proceeds; shore authorities are alerted and over the next hour the shore support facility should become fully effective.

Around the emergency control centre the OIM is pleased to see assessed information beginning to take shape. Unfortunately one of the key presentation aids is being set up by a man he knows has not done it before. He resists the temptation to give this man some help, and mentions the matter to his Maintenance Supervisor - who responds as if he has not quite woken up fully. The assessment of fire and explosion risks looks as if it will involve some careful consideration of possibly quite complex safety-case factors. The OIM looks round for the Construction Supervisor and draws him into a quick discussion with the Maintenance Supervisor. What seems to be the damage from the explosion?

The Maintenance Supervisor is talking to the On-scene Commander - a good man; Questions race through the OIM's mind. Is the fire containable?; what inventory of flammables and toxic substances is involved, and what does this imply?; what options have we to help the men who may be trapped (the main muster is still incomplete)?; what is the hazard to the Temporary Refuge (TR)?; what is the hazard to the platform structure?. The OIM invites the Maintenance Supervisor to look into the integrity of escape means and routes to lifeboats and, of course, the helideck. Then he has a very quick few words with the Production Superintendent.

You will reflect that there is time, apparently, to hold this sort of brisk succinct consultation. Emergencies generally permit this, in that even the worst of them (including Piper Alpha, for instance) comprise peaks and troughs of activity. It is essential that the command team can identify the troughs as vital opportunities in the process of regaining the initiative. Needless to say this is easier said than done, especially when fear and stress stalk the mind.

The OIM calls once again for attention, reviews the position (already covered with his principals), puts a couple of suggestion to the Maintenance Supervisor, and once again makes a PA announcement. It is seven minutes since the explosion - it seems like an hour to the OIM, but he fears it may seem like little more than a minute to the Maintenance Supervisor - who, he is concerned to see, still looks a bit off colour.

Strategy formation: it is vital to survival

A strategy is a key step to the longer-term objectives and direction. Consider how this emergency might progress in terms of the questions we have asked ourselves. The explosion *may* have damaged the fire wall between the main production module (where the explosion occurred) and the adjacent gas separation module, but we have not yet gained sufficiently close access to be sure one way or the other. The fire should be containable provided the inventory is properly isolated; but there are unexpected indications of residual pressure - have all valves closed correctly? We are shifting clear methanol drums stored above the fire, but temperatures are rising and there is concern for those involved - can we complete this task? There is no immediate threat to the Temporary Refuge. We are concerned as to whether we can gain sufficient control of the fire (deluges, and a foam blanket under the separator bottom run-offs) before risk of separator rupture becomes unacceptable and the fire teams must withdraw. But there is increasing evidence that there are three casualties in the production module - none of these were present at muster. The helideck is usable, the quantities of thick black smoke blowing clear. We hear that one Coastguard helicopter will be in the field in just over an hour, but it seems then we must expect another forty minutes before anything else arrives, including specialist medical assistance. Launching lifeboats is possible, but it is overcast and apparently very dark. Our excellent Standby Vessel, which is about to engage in some carefully-directed support fire-fighting, will be joined in 35 minutes by a less-capable sister from an adjacent platform. There is a supply vessel also with us.

The OIM must force himself away from the detail. Everyone proclaims this mantra. It is however not at all easy for a person dedicated to production efficiency most days of working life to detach from the beloved plant and become a high-level strategist addressing wider and generally more demanding problems. Time after time, in training, we watch OIMs under pressure become drawn inexorably back into the tactical fray of damage control and the intricacies of personnel accounting. However well-intended, this sadly is too nearly the equivalent of fiddling while Rome burns.

Our OIM, happily, has no such leanings. Whilst keeping intuitive ears and senses tuned to what is going on around him, and keeping an eye on the Maintenance Supervisor (who at least seems to be getting into his stride), he turns his mind firmly to the future.

Naturalistic decision-making can obviously help here, but the factors are many and their manifestations complex. This is no time to jump to conclusions. To achieve balanced objectives some measured development of options and their analysis is essential. The OIM may be beginning to consider whether the fire escalation is on a scale that makes it important only to the extent that it influences the overall strategy for the saving of life in what are becoming life-threatening circumstances for everyone on the platform. What attention can he afford to spare for the trapped men? Can he place further men at risk, now, on their account?. Is there some imaginative option that could help them?.

With intermissions, the OIM devotes some three minutes to developing two prime options. Sensing a slight lull, he detaches the Maintenance Supervisor and takes him into an adjacent office with the Services Supervisor. The Services Supervisor (a steady man) endorses his thinking; the Maintenance Supervisor just nods. The OIM sends the latter back and invite the Services Supervisor to work up the plans. Returning then to the emergency command centre the OIM senses a deterioration. Information is not coming along well; he has to chase for a clear picture, losing a minute. Another review of immediate options, objectives stated and a PA broadcast to keep spirits up whilst warning for possibilities of evacuation, it is time to speak to the asset manager.

And so on. As you can see the incident may pass through several key changes of direction. These may include: escalation; effective control; buying time before escalation proceeds.

The emergency commander must be able to identify swiftly such swerves in direction and adjust strategy accordingly. At each stage, there are decisions to be made and we know already that the commander under time pressure is likely to pick the first workable option that occurs. There is a need to bear in mind always the worst credible scenario, whilst not "taking counsel of one's worst fears", and to test and question the current actions and decisions points to search for any anomalies which might indicate that a key decision needs to be revised. This of course introduces another level of analytical thinking.

As always with emergency management, it is easy to say "how obvious". But it is when the factors of time pressure and incomplete information are coupled with activities beyond the operational envelope that the true complexity and risk of disorientation emerges. I hope I have helped to bring alive aspects of this process. Good emergency simulation exercises do help enormously. But there is no substitute for reality in terms of actual experience. Continuing still on the theme of creating direction, I would now like to touch on these two factors.

Time pressure and incomplete information are an uncomfortable but often unavoidable fellow travellers; they heat decision making, and are stressors.

Time pressure and incomplete information are both stressors. Under stress, as we have seen from our Maintenance Supervisor, an individual's behaviour patterns may alter. To begin with the outer physical signs are small and difficult to identify but, as levels of stress increase, obvious physical manifestations take place and the capability of an individual to function effectively is impaired. The symptoms of stress vary with each individual - each of us has strengths and weaknesses. The experience of the submarine command 'Perisher' course provides rich prime evidence here. From observation of industry managers both onshore and offshore, we have identified similar behaviour traits during emergency training programmes and simulation exercises. Let us place these illustratively in the first person.

- We may see reality skewed, as through a prism - deny the problems we face, and see instead reflections that we find more malleable, and therefore preferable.
- We may see a confused mass of data, deducing nothing of value. We settle for the answer we last used or the one we are offered, or we just guess.
- We deny there is a problem, and we deal with trivia. We act with no clear objective, no focus, no priority.
- We do not understand the information presented. We misinterpret, draw wrong conclusions, followed by wrong judgements.
- We panic. In clear sight of reality and its awful implications, we freeze, do nothing or become emotional.
- We interpret the information correctly and make good decisions, but organise resources ineffectively and fail to control the situation.

In the Royal Navy's case, candidate submarine commanders, already well trained and experienced in emergency command, are subjected systematically to extremes of pressure. Some 25% fail to qualify as competent. We do not advise that this level of pressure is useful, let alone necessary, for industry management training. We would however I think agree that industry emergency commanders may, in reality, be subjected to pressures under which their performance deteriorates. Ultimately this will be true of any individual, since everyone of us has a limit of resistance and endurance. We shall return to this point later, and consider to what extent it should be a concern for industry managers and what they can do about it.

Let us now leave our example in the midst of the fight for a satisfactory outcome, and look at one or two of the key mechanisms in more detail.

There is a crucial relationship between strategy, decision-making, time pressure and the operational envelope.

We return to the research on fire team leaders (Klein). Interestingly, it was found that the fire team leaders coped well under the normal levels of pressure encountered in their work. This in part could be due to the fact that, when studied, they were largely within their operational envelope. It could also reflect that the structured and natural selection processes which have taken place over a period of time have worked well. Certainly, being within one's operational envelope enhances the ability to cope under the pressure of time constraints and information shortage. Outside the operational envelope, people must - as we have seen - work things out for themselves, although we must hope they will not be called upon to do so entirely from first principles. If the speed of their analytical processes, under pressure, is insufficient for the circumstances, the performance of an individual will be affected.

Observation of industry major hazard managers shows that the ability to function outside the operational envelope is developed through a process of natural selection until, at senior management level, one of the key functions is an ability to develop an accurate strategic overview and 'ask the right questions'. Therefore our chances of finding such people in positions of responsibility for emergency command prove, in practice, to be understandably high. This is indeed demonstrated by the majority of managers whom we have trained and assessed.

Take a closer look at time management

As an emergency progresses, more information becomes available and decisions become easier to make. In parallel, however, options are closing. It is therefore extremely important to manage the time-base of an emergency - i.e. to have a clear picture of the timescales within which actions may necessarily have to take place and to provide the basis for a forward plan. Various techniques of action plotting are available. Recalling our scenario, the OIM must develop an appreciation of the critical time windows available. For example, a reverse critical path analysis can be done quite quickly:

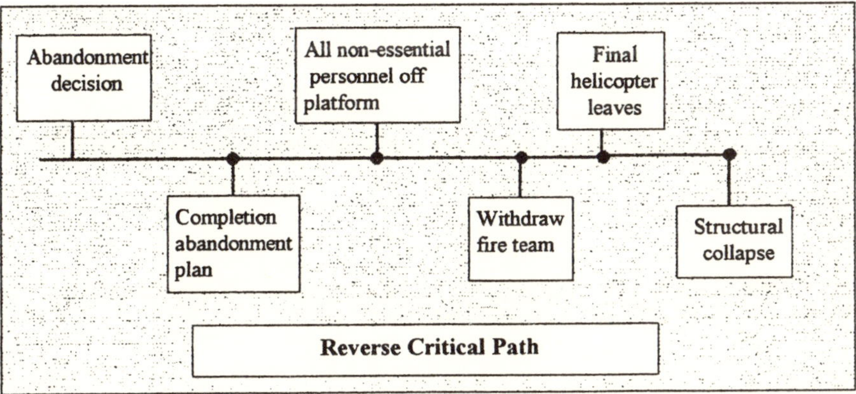

Figure 3.

Vitally, if self-evidently, time is liable to be at a premium and the emergency commander must make best use of it. We have seen that there will be opportunities during the emergency for thought and consideration of options and these must not be squandered. An example of less than ideal behaviour was an emergency commander who effectively turned into the telephone operator, dutifully obeying the orders of his command team. Yet we have had it suggested seriously that a prime function for the OIM is to keep his asset manager ashore constantly informed personally.

Team management and leadership are potentially winning commodities

The qualities of a leader are a subject in themselves and go beyond the confines of this paper. I will restrict myself to presenting a few basic concepts in support of the general theme of emergency management. One of the pre-requisites of effective emergency management is the structuring and training of an effective supporting team. For any emergency configuration it is useful to perform an elementary task analysis. Returning to our case study, one suitable emergency team configuration might look like this:

- OIM - overall command
- Production liaison - control of production and processing plant, technical advice and diagnosis, deployment of process team. May be deputy and key advisor to the OIM.
- On-scene and Maintenance/Engineering - link to and direction of incident; also takes non-processing issues. May be deputy and key advisor to the OIM.
- Logistics - Control of helicopter and vessel movements, liaison with external agencies.
- Personnel mustering - muster co-ordination, personnel movement, abandonment planning.
- Log-keeper

The number of people involved could of course be more or less than this depending on the characteristics of the installation.

Consider the key points in constructing an emergency team:

- There must be a balance between totally overloading a single individual with information-handling duties and creating, at the other extreme, too complex a matrix of interfaces.
- Information handling and presentation systems must reflect the specific needs of the team and the demands placed by the nature of the installation.
- Team members should be trained to observe and support each other, including support of the emergency commander at difficult moments.

Project into the future

A self-evident truth, one of the many with which emergency management abounds, is that one can do nothing about the past, one can act in the present, but one can influence the future. The emergency manager's role is to ensure a successful outcome in the future. In major hazard industry this usually equates to the saving of life as a first priority and saving of plant the second. Observation of the attention focus of emergency commanders under training has shown the following:

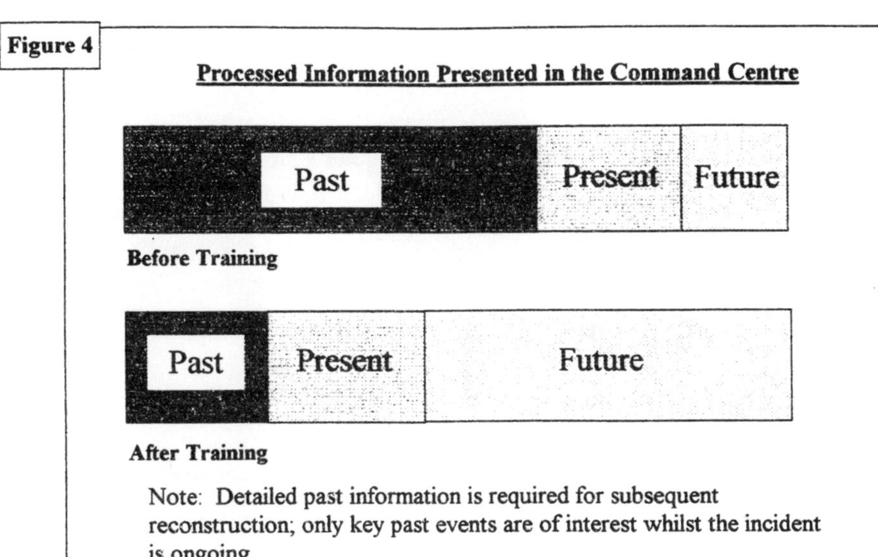

With training, successful emergency managers develop a more useful focus of attention as they become more effective. They begin to consider and balance their present decisions against future possibilities. Both logic and practical observation support the conclusion that this skill is one of the last to be learned by an emergency commander. While an individual is seeking to grasp the basics, the finer points are elusive. In due course, when the basic skills are assimilated thoroughly, an individual can move on to complete the range of core skills.

TRAINING OF EMERGENCY MANAGERS

As we have shown, emergency managers require a broad base of skills. These include:

- Capability to manage under pressure;
- Capability to operate outside the operational envelope. The broader the range of operational experience, the less likely is the emergency commander to find himself having to operate outside the personal operational envelope.
- Ability in emergency team command; leadership;
- Skills in the basics of emergency management - logistics; personnel accounting and movement; evacuation plans; communication and information processing;
- Skills in major hazard management - process plant; drilling; technical diagnosis.
- Knowledge of major hazards: fire and explosion; chemical; radiation;
- Knowledge of the facility itself.

In training personnel across such a broad front, it is important to take a broad view. There are some key features of emergency command training which I would like to cover below:

Emergency management training is an intellectual not a physical exercise

To illustrate this point, it is perhaps helpful to consider the levels of emergency command: These line up broadly with conventional management terminology in the following way:

On-scene commander; fire team leader	First line management
Production superintendent; shift charge engineer; shift manager	Second line management
OIM (large platforms); Station Manager	Senior management

From first line management through to senior management levels, there is an increase in the strategic intellectual activities at the expense of the tactical and practical, although contact with the practical level should always be maintained.

Concerning stress, we have seen that it is not necessary to subject people to extremes of pressure and fear in order to produce an adequate vehicle for training "outside the operational envelope". Indeed it is generally inadvisable. Training can however usefully take place in a physical environment, offering a gentle degree of stimulation and excitement, but offering predominantly the opportunity to develop techniques for the management of critical fast-moving situations. Let it be understood that this is primarily an intellectual, not a physical activity.

Depth of training required

On the personal competence ladder, we start at a level of unconscious incompetence: we don't know what we don't know. We develop an awareness which takes us to conscious incompetence: we know what we don't know. We develop a conscious competence: so long as we think about what we are doing, we can do it. Ultimately we may arrive at a stage of unconscious competence: here we act correctly almost without thinking. One of the objectives of OCTO's emergency management training is for people to attain a state of unconscious competence in most of the basis skills, so that they can begin to focus on the strategic, time-based and decision-making skills and achieve at least a stage of conscious competence in these areas. There is a real danger, too frequently demonstrated, that people stop at the level of conscious competence in the basics and fail to address the core issues. As a result, when the pressure is on, the emergency command structure breaks down, with potentially disastrous consequences. Note that a high level conscious competence may be all we should expect reasonably to achieve overall as managers of a complex plant. Unconscious competence can smack of over-confidence and, ultimately, superficiality.

Extensive experience of training in command competence shows that for many people the goal is intangible. A successful technique has been to show individuals by experience what they can achieve, whereby they can create their own personal image and understanding of command through practical experience under expert guidance. The key is to relate theoretical, intellectual training to practical contexts of emergency exercises, emergency simulations and practical command training. Basically this means:

Don't:

- Teach dry management, command and decision-making models in isolation.
- Train individuals in the basic elements of emergency management without also giving them training in putting all the elements together into a coherent and effective whole.
- Re-inforce bad-practice by giving either no feedback in simulation exercises or only generic feedback.

Do:

- Use trainers who are expert in the field and able to follow the cognitive processes of the emergency manager who can therefore help to diagnose individual difficulties and coach individuals towards a solution.
- Perform on-line coaching through simulation exercises - whether on or off-site, incorporating immediate feedback based on specific observed behaviour.

Interestingly the recent research by Klein in the United States into command team decision making is reinforcing this experience.

Finally, in case anyone still harbours doubts as to the realism or attainability of high standards, I would like to conclude with some myths of emergency management training (Fig. 6 - final sheet of text).

References

Charlton, Commander D. Royal Navy (1992). Training and Assessing Submarine Commanders on the Perishers' Course. Paper presented at the First Offshore Installation Management Conference, The Robert Gordon Institute of Technology, Aberdeen.

Cullen, The Hon. Lord (1990). The Public Enquiry into the Piper Alpha Disaster. Vols. I and II. London: HMSO. CM1310.

HSE (1992). A guide to the offshore installations (safety case) regulations (1992). London: HMSO.

Klein,G. (1993). A recognition-primed decision (RPD) model of rapid decision making. In G. Klein, J. Orasanau, R. Calderwood, & C. Zsambok (Eds.), Decision making in action: Models and Methods (pp.138-147). Norwood, NJ: Ablex.

Larken (1992). Command as a feature of crisis prevention and crisis management in the offshore industry. Institution of Marine Engineers Conference on Offshore Safety: Protection of Life and the Environment.

Larken (1993). Prevention and control of emergencies - the senior manager and command development". Twelfth Leith International Conference, Edinburgh.

Larken (1994). "Selecting and training emergency incident controllers". The AEA Technology Consultancy Services SRD Association Conference.

Orasanau, J. (1994). Decision making in action: Meeting the challenge of emergency events. Paper presented at the Third Offshore Installation Management Conference, The Robert Gordon University, Aberdeen.

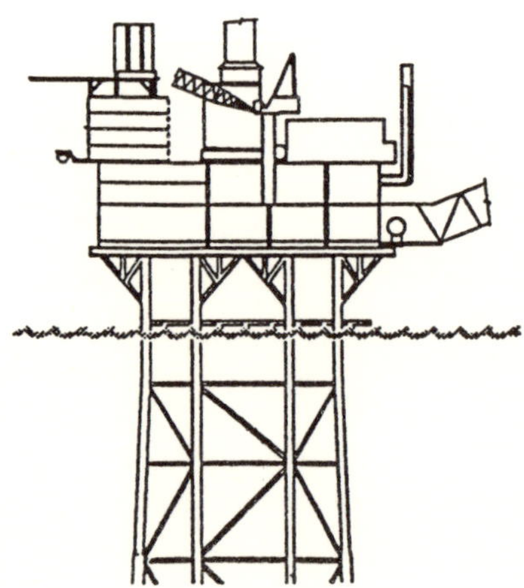

Figure 1

Plant Status:

- Shutdown
- Blowdown?

Fire/explosion Risks

- Duration?
- Escalation?
- Consequences?
- Options to control and mitigate?

Personnel Accounting

Muster status?

Casualty status and prognosis?

Exposure to hazard?

Options to protect those at risk?

Logistics

Vessels

Helicopters

Search and Rescue

Medical experts

Resources - Options to deploy

Fire teams

Medical team

Production team

Crane operator

Helideck crew

Figure 2

MYTHS OF EMERGENCY MANAGEMENT TRAINING

Emergency management training is expensive.

The key to cost-effective training is to include high quality training in routine emergency exercises and simulations and always to aim as close as possible to reality. Emergency training is much cheaper than dealing with the aftermath of ineffective emergency management.

Experienced industry managers don't need training.

Ask the managers. From my experience, people who know they will be in the hot seat welcome all the good advice they can get.

It is possible to write procedures to deal with any emergency - so why train people to work without them?

Extensive experience in the military and in industry situations shows that emergencies very quickly move beyond the limits of prescribed procedures. Any assumption that they will not is foolhardy. It also collapses under the evidence of accident analysis.

Industry needs managers in place to prevent emergencies; the skills of emergency management are not compatible.

Of course the primary role of a major hazard manager is to prevent an emergency. Experience shows however that the ability to command in an emergency almost always coexists with abilities in general and safety management and indeed enhances the ability to operate in fast-moving, high pressure circumstances.

It is impossible to assess the ability of people to perform under pressure

It is difficult, and it needs a high standard of expert and specialist observation and experience in emergency management. It is however eminently possible and normally extremely useful for individuals to learn constructively about their own strengths and weaknesses under pressure.

There is no point in training an industry manager to manage a difficult emergency because, unlike the military who do it all the time, industry managers do it so rarely that they are bound to fail in practice.

One view on this is the general public - who expect major hazard industries to be able to deal with the consequences of their mistakes without major repercussions on the workforce or the general public. Lord Cullen makes this point in his report into the Piper Alpha disaster. Another point is that industry managers start off with some excellent characteristics and can indeed be trained to a very good standard. Also a reasonable amount of practice and some carefully targeted refresher training will keep them there. Good standards can be maintained with a well-considered emergency exercise programme which is probably no more onerous than that currently in place.

Figure 5